Fluids

Author
Nora L. Alexander

Program Consultant
Marietta (Mars) Bloch

Nelson
Thomson Learning

Australia • Canada • Denmark • Japan • Mexico • New Zealand • Philippines
Puerto Rico • Singapore • South Africa • Spain • United Kingdom • United States

Contents

 Important safety information

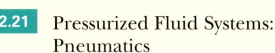 Record observations or data

① Refer to numbered section in *Nelson Science & Technology 7/8 Skills Handbook*

Unit 2 Overview

Fluids cover Earth's surface. You breathe them and drink them. They flow through your body and in your home. Many machines and devices use fluids, from dentists' chairs to propane barbecues. The use and investigation of fluids affect our lives.

What are fluids? How are they used? What advances in technology have been made from understanding fluids?

Properties of Fluids

Air, engine oil, propane, and corn syrup are all fluids. What do they have in common? How are they different? As we investigate fluids, we find that they have different properties that we can describe: viscosity, buoyancy, density, and pressure.

You will be able to:

- predict how temperature affects the flow rate of liquids

- compare the viscosities and densities of various substances, and understand the relationship between the two properties

- describe the difference between mass and weight

- measure the density of liquids with an instrument you have designed and constructed

- explain how buoyancy and gravity are related

- use the particle theory of matter to compare the densities of solids, liquids, and gases, and to describe the effects of temperature on fluids

- compare what happens to different fluids when an external pressure is applied or when the temperature is changed

The Use of Fluids

A fluid's properties may pose some challenges, such as getting the ketchup out of its bottle! But we can also make the properties of fluids work for us. Fluids for example, fuel the airplane below and allow the pilot to move the flaps on the airplane's wings.

You will be able to:

- plan and conduct an investigation to compare the densities of liquids and solids

- explain how the buoyant force on an object can be altered

- explain how the properties of fluids are used in the design of technical innovations

- design and build a model of a device that uses hydraulics or pneumatics

- compare how liquids and gases transmit force in hydraulic and pneumatic systems

▲

Fluids and Living Organisms

Fluids have a wide range of properties. By looking at nature—the shape of a whale and the stickiness of tree sap—we can learn a lot about fluids that will help us to improve design technology. At the same time, investigating fluids and their properties in artificial systems can help us understand the natural world better.

You will be able to:

- compare how fluids function in living things and manufactured devices

- explain how the study of hydraulics and pneumatics increases understanding of fluids in the human body

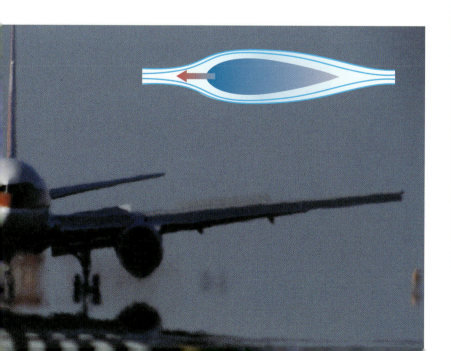

Design Challenge

You will be able to...

demonstrate your learning by completing a Design Challenge.

Devices That Use the Properties of Fluids

You will design and build a model of a device that uses the properties of a fluid to perform a certain function. Your model will help you to understand the relationship between the properties of fluids and their use in machines and devices.

In this unit you will be able to design and build:

1 **A Pneumatically or Hydraulically Operated Safety Hinge**

Design and build a model of a door that opens or closes using pneumatics or hydraulics.

2 **A Boat Navigation Lock**

Design and build a model of a navigation lock that allows boats to travel between waterways of different levels.

3 **A Fish-Tank Cleaner**

Design and build a fish-tank cleaning device that will raise or lower itself in water.

To start your Design Challenge, see page 54.

Record your thoughts and design ideas for the Challenge when you see

Design Challenge

Getting Started

Fluids in Our Lives

1 Name a fluid. You probably think of one like water, and you are right. Liquids are fluids, but so are gases such as the air we breathe. Although fluids have properties they share with other substances—they take up space and are made of matter—fluids also have unique properties. Liquids can be thick or thin. Some things float in fluids, and others sink.

In what other ways are we concerned with the movement of fluids? What do you think are some of the properties of fluids?

2 Fluids help to make our lives easier. Engineers harness the energy of water to generate electricity. As air rushes into a vacuum cleaner, it carries dirt and dust along with it.

A machine called a tree spade enables us to transplant very large trees. The blades are forced open and then down by the movement of oil within cylinders in the spade.

What other problems can we solve by investigating and using fluids? What kinds of devices and machines do we make that use fluids?

Reflecting

Think about the questions in ❶,❷,❸. What other questions do you have about fluids? As you progress through this unit, reflect on your answers and revise them based on what you have learned.

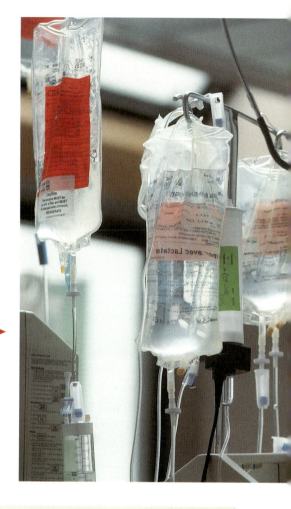

3 The health of Earth's fluid systems, including its waterways and atmosphere, is necessary to sustain our lives and the lives of future generations. Fluids such as blood and air keep us alive. Intravenous fluids are delivered to hospital patients when needed. Is the way fluids function in living things the same as the way they function in systems we create? How do changes in fluids affect living things?

Try This | Where Are Fluids in Your Life? 9E

Think about the fluids in your life and the devices that use them.

- Work with a small group. Each member should have about 20 pieces of paper, each approximately 6 cm × 6 cm. On each piece of paper, write down any fluid you can think of or any instrument or machine that uses a fluid. Use only one piece of paper for each item.

1. How do you decide what things are fluids and what things aren't?

- When each group member has used up all 20 pieces, combine your group's papers into one pile. Separate the words into categories.

2. What categories are you using?

- Glue the words in their categories onto a poster-sized piece of paper. Use markers to give each category a heading and to draw arrows between words that are connected. You are making a word map. Draw illustrations where appropriate. Write the word *fluids* somewhere on the word map.

3. Compare your group's word map to another's. Are the categories similar? Why might there be differences?

A Close-Up Look at Fluid Flow

Fluids are substances that flow (**Figure 1**). Water flows from the tap when you wash your hands; you see it flowing in a stream. But liquids are not the only substances that flow. What flows past your face when you coast your bicycle down a hill? What carries dandelion seed fluff around the neighbourhood in the spring? Air, which is a gas, also flows. Both gases and liquids are fluids.

Systems involving moving fluids are a concern for people in many professions and fields. (See **Figure 2**.) How will a tower withstand a gusty wind? How will deposits on artery walls affect the flow of blood? Is the inside of a pipe smooth enough to enable natural gas to flow safely? How is an airplane affected by different kinds of airflow? The concern with fluid flow exists both when a fluid is moving and when an object is moving through the fluid. How quickly a fluid flows in a given amount of time is called its **flow rate**.

Systems involving moving fluids are said to be **dynamic**. **Aerodynamics** refers to air (gas) moving around solid objects. **Hydrodynamics** refers to the motion of liquids (usually water) around solid objects.

Figure 1

Flow tests are conducted on fire hydrants to ensure there will be enough water in an emergency.

Try This   Determining Flow Rate

You can find the rate at which the water flows from the tap in your classroom or outside the school.

- Calibrate a large bucket at the 10-L mark by measuring water into it using a 1-L container, such as a plastic beaker. Mark the 10-L water level with a black waterproof marker.
- Place a mark on the handle of the tap. Determine how many turns of the tap (how many times the black mark rotates) are required to open the tap fully.

1. When the tap is opened fully, how long does it take to fill the bucket to the 10-L mark?

2. When the tap is opened halfway, how long does it take to fill the bucket to the 10-L mark?

3. How does the result you obtained in question 2 compare to the result in question 1? Is this what you expected to happen? Explain.

- The volume of liquid that flows in a second is called its flow rate. Calculate the flow rate of the tap in litres per second.

Solids That Seem to Flow

A fine powder, consisting of a very large number of tiny solid pieces, can be poured from one container into another. (See **Figure 2**.) But have you ever seen water form a heap, as flour does when you pour it? Can you make a pile of milk, as you can of sand or wheat? The answer, of course, is no: only solids can be piled in a heap. Liquids and gases fill whatever container they are in.

Figure 2

Here is a material that is not a gas or a liquid but appears to flow. Flour, sand, and wheat all seem to flow. Why are they not considered to be fluids?

Explaining Flow Using the Particle Theory

The **particle theory** of matter provides a model to help us understand the differences between fluids and solids. It also aids in understanding and predicting fluid behaviour.

The particle theory states that
- all matter is composed of particles;
- particles are in constant motion;
- there are forces of attraction among particles.

When particles are close together and moving slowly, the forces of attraction are strong. The particles in a solid are so close together and their forces of attraction are so strong that they cannot flow past one another. In liquids the particles are moving more rapidly, so forces of attraction between them are not as strong. Because the particles are not locked in a fixed arrangement, they move a little farther apart and can slide over one another. This explains why liquids are capable of flowing. In gases the particles are so far apart from each other and the forces of attraction are so weak that the particles can move independently of each other. As a result, gases flow very easily.

Understanding Concepts

1. Using the particle theory, explain why solids do not flow.

Making Connections

2. How is the flow of air used in transportation?

Exploring

3. Investigate how a water-saving (4A) shower head works.

Reflecting

4. Take another look at the word map you prepared in the Getting Started Try This activity. Are any of your examples solids that seem to flow? Should they remain on the map?

Design Challenge

You are using a fluid in your Design Challenge. How will you consider the flow of that fluid during the design and testing of your model?

Fluid Flow Around Objects

The shape of an object determines how fluids flow around it. Consider the flow of water in a river. A deep river, with steep banks and no obstacles, flows fast and smoothly. The water travels in straight or almost straight lines. This is known as **laminar flow**. Now imagine a shallow river, with irregular rocks breaking the surface. The water will be broken and choppy—unable to flow in straight lines. This is called **turbulent flow**, which may result in rapids, eddies, and whirlpools.

The same thing occurs with gases in motion. As moving air encounters objects such as buildings or trees, the flow becomes turbulent. **Figure 1** illustrates laminar and turbulent flow.

Shapes that produce a laminar flow have less air or water resistance than shapes that produce a turbulent flow (**Figure 2**). Resistance is referred to as **drag**. For cars and airplanes travelling at high speeds, less drag means better fuel consumption and less wind noise. Shapes that create a laminar flow are said to be **streamlined** or aerodynamic. (See **Figure 3**.)

A fluid moving relative to an object experiences resistance as its particles slam into the object. Water flowing under a bridge meets resistance as it passes the piers. Airplanes encounter the resistance of air when they are flying. Objects falling through the air are slowed down because of air resistance.

Figure 1

a Laminar flow around an object

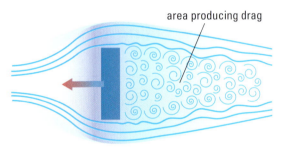

area producing drag

b Turbulent flow around an object

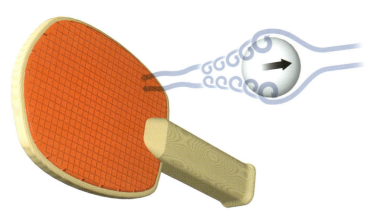

Figure 2

Turbulent and laminar flow can be used to control movement and direction. Sports balls are designed with this in mind. The airflow around this ball becomes turbulent at the top and bottom of the ball. This helps to slow the ball.

Wind Tunnels: A Closer Look at Gas Flow

Canadian Wallace Rupert Turnbull is credited with building Canada's first wind tunnel in 1902. He conducted experiments in the tunnel to test his propeller inventions.

A wind tunnel has a propeller at one end that propels air into it. Smoke is often added to make the flow of air visible.

Wind tunnels are widely used today. (See **Figure 4**.) Engineers use them to test the airflow around wings of aircraft and investigate how ice on wings affects airflow. Vehicles are examined in wind tunnels to determine how streamlined they are. By placing precisely designed scale models of tall buildings, bridges, and towers in wind tunnels, engineers can examine how the structures are affected by high winds.

Figure 4

Figure 3

The bodies of whales and dolphins are streamlined for decreased water resistance. This is achieved by an elongated shape with no narrowing at the neck, no protruding parts, and smooth skin. The tail fluke produces a more laminar flow of water around the body. Notice how a side view of one-half of the tail fluke resembles an aircraft wing.

Understanding Concepts

1. Make a chart with two headings: Laminar Flow and Turbulent Flow. List some examples of each type of flow.

Making Connections

2. Why might a car manufacturer change the shape of side mirrors on a particular model?

3. How would deposits on the walls of arteries affect the flow of blood moving through the body?

4. Would you prefer turbulent flow or laminar flow around a racing bicycle? Which would you like for whitewater rafting?

Exploring

5. Research the importance of air (4A) resistance in your favourite high-speed sport.

Reflecting

6. Why do scientists study airflow?

Viscosity: A Property of Fluids

Have you ever tried to pour ketchup out of a brand new bottle? It takes a lot of force to start the ketchup flowing. (See **Figure 1**.) Very little force is required to start maple syrup flowing, as **Figure 2** shows. That's because maple syrup is much thinner and has less resistance to flowing than ketchup. **Viscosity** is the term for the resistance that a fluid has to flowing and movement. The particle theory helps us to understand that this resistance is due to the forces of attraction among particles. The stronger the attraction among the particles, the greater the resistance of the particles to flowing past one another. Different substances are composed of different particles and have different forces of attraction. This helps to explain why fluids can have different viscosities.

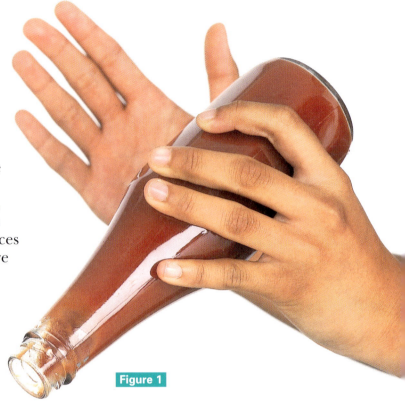

Figure 1

Ketchup is thick and has a high viscosity. Maple syrup is less viscous (has a lower viscosity) than ketchup. Water is a thin, runny liquid with a low viscosity. Maple syrup is more viscous than water, but less viscous than ketchup. Water has less resistance to flowing and movement than maple syrup.

When fluids are stationary, viscosity is not a concern. However, when a fluid is moving, or when something is moving through a fluid, the property of viscosity can be very important.

Figure 2

Measuring Viscosity

We often need to know how quickly or how slowly a fluid flows. If you tipped a water pitcher as quickly as a salad dressing bottle, you might soon find a puddle of water on the table! You handle the two fluids differently because you know that they have different viscosities.

We might use the words *thick* and *thin* to describe viscosity, but these words do not give enough information about this property. We need some way of measuring viscosity quantitatively. One method involves timing how quickly a solid, such as a pearl-sized ball, falls through a column of the liquid. Another method times how long it takes a liquid at a certain temperature to flow into a small pot.

An instrument that measures viscosity is called a **viscometer**.

Design Challenge

How viscous is the fluid you are using in your Design Challenge? What effect does its viscosity have on your design?

Understanding Concepts

1. Molasses has a high viscosity. Explain what this statement means.

2. How does the thickness of a fluid compare to its viscosity? Give an example.

Making Connections

3. Research an industry in which viscosity testing is important. Present your findings as a poster. (4A)

Exploring

4. Research how soap affects the viscosity of water.

5. Plan an investigation to help you rank the following fluids in order, from least viscous to most viscous: shampoo, melted chocolate, cooking oil, air, and cola. (2E)

Reflecting

6. What does the viscosity of a fluid tell you about its flow rate?

Try This Let's Examine Moving Fluids

Figure 3

Figure 4

- Fill a small, clear plastic bottle with a tight-fitting lid with corn syrup, leaving a 3-cm space at the top. Fill another with water, leaving a 3-cm space at the top. Add a small, identical amount of paper confetti to each bottle. Fasten the lids securely (**Figure 3**).

As each bottle moves, carefully observe the confetti.

1. How does the movement of the confetti in the two liquids compare?

- Using a board at least 110 cm long, build a ramp with a low incline. (See **Figure 4**.) Roll the bottles, one at a time, down the ramp. Observe the movement of the confetti.

2. Sketch your observations.

3. What conclusions can you make about the movement of the two liquids?

4. Why do you think you were told to add confetti to the bottles? What might the confetti represent?

Liquids Can Be Thick or Thin

Have you ever put the corn syrup bottle in the fridge instead of the cupboard? What happens when you try to pour the syrup? You have a problem! How could you make it easier to pour? What happens to the syrup when it lands on your hot pancakes? All these questions involve viscosity and temperature. Are these two properties related?

Question

Does temperature affect the viscosity of oil? (See **Figure 1**.) If so, how? How can this change be measured?

Hypothesis

2C Predict the change in viscosity as heat is added to or removed from a sample of oil.

Experimental Design

In this investigation you will explore what happens to viscosity when a fluid is heated or cooled. You will use a homemade viscometer to measure the flow rate of oil at three temperatures. Flow rate is a measure of viscosity.

1 Copy the observation chart in **Table 1**.

Materials

- apron
- safety goggles
- water
- 150-mL foam cup
- beaker
- wax pencil
- retort stand
- ring clamp (for foam cup)
- small metal skewer
- ruler
- stopwatch
- 500-mL beaker
- 80 mL of cooking oil at three temperatures: 20–24°C, 5–8°C, and 45–50°C
- thermometer
- tissues
- hot plate

Procedure

2 Using 70 mL of water as a guide, mark the 70-mL line on your beaker.
- Empty the container and wipe it dry.

3 Use the skewer to poke a hole in the bottom of the foam cup.
- Attach the ring clamp 30 cm from the table top. Sit the foam cup in the ring.
- Put the beaker underneath the cup.

4 Measure the temperature of the 20–24°C oil.

✎ (a) Record the actual temperature in your chart.

- Wipe the thermometer.
- Tightly cover the hole in the cup with a finger and pour the sample into it.
- Remove your finger and immediately start timing.

5 Stop when the oil reaches the mark on the beaker.

✎ (a) Record this time, calculate, and record the flow rate. Use the formula.

$$\text{flow rate} = \frac{\text{volume of fluid (mL)}}{\text{time (s)}}$$

- Allow the cup to drain. Empty the beaker according to your teacher's instructions. Wipe out both containers.

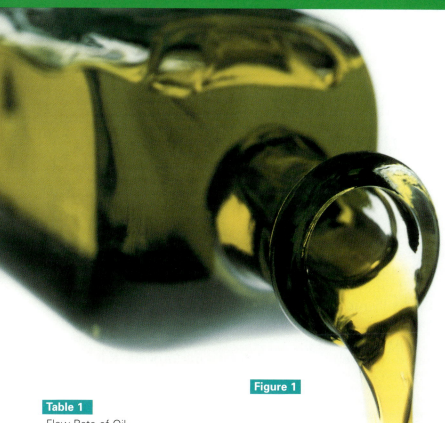

Figure 1

Table 1

Flow Rate of Oil

Oil Temp.	20–24°C	5–8°C	45–50°C
Time	?	?	?
Flow Rate	?	?	?
Appearance	?	?	?

Making Connections

1. From your graph, predict what flow rates you would expect for oil at 12°C and at 100°C. Explain your reasoning.

   🛑 Do NOT try heating oil to 100°C.

2. How would you solve the corn syrup problem mentioned in the introduction to this Investigation?

3. What household products could require viscosity testing during manufacturing?

Reflecting

4. Why were you given 80 mL of oil, but asked to record the time for 70 mL to flow?

6 Repeat steps 4 and 5 with the 5–8°C and 45–50°C oil samples.

(a) Describe the appearance of the oil at these temperatures.

7 🄬 Make a graph of your results. Put temperature on the *x*-axis and flow rate on the *y*-axis. Give your graph a title.

Analysis

8 Analyze your results by answering the following questions.

(a) At which temperature is the oil most viscous? Give a qualitative and quantitative description of the oil that supports your answer.

(b) At which temperature did the oil have the highest flow rate? What does this tell you about the viscosity of the oil at that temperature?

(c) What relationship exists between the temperature of the oil and its flow rate?

(d) Why did you wipe out the beaker after step 5 each time?

9 🅰 Write a formal lab report for this investigation.

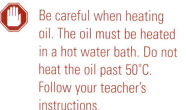

Be careful when heating oil. The oil must be heated in a hot water bath. Do not heat the oil past 50°C. Follow your teacher's instructions.

Remain standing as you perform this investigation.

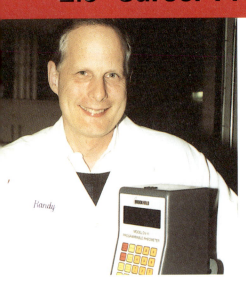

Viscosity and the Chocolate Factory

Randy Droniuk is a food scientist. His career is the envy of chocolate lovers everywhere. Randy is one of the many people working every day in a factory to ensure the quality of chocolate products are delicious. He runs tests during the chocolate-making process, and he researches how to improve the process. "I enjoy the variety of work involved in my job," says Randy. "It is really nice to work on anything that involves a better quality product for our customer. I also love to eat chocolate. My favourite is dark chocolate."

To test and research the chocolate-making process, Randy needs to understand the property of viscosity and how it applies to liquid chocolate. "Viscosity testing is very important in this industry. We want to ensure your favourite chocolate bar is of the same high quality from batch to batch. Both temperature and ingredients greatly influence viscosity. By running regular tests, we can produce a dependable product."

Figure 1

Some chocolate bars are made in moulds. Liquid chocolate, at approximately 30°C, is poured into moulds that resemble ice cube trays. A vibrator settles the chocolate into the moulds. Filled moulds are cooled slowly, and the chocolate solidifies. The moulds are turned over and out falls a chocolate bar ready to be packaged.

The Importance of Viscosity

Not all chocolate bars are the same. Neither is the chocolate that goes into the many different varieties of these treats.

Scientists in a chocolate factory measure the flow rates of liquid chocolate. A different viscosity is required for moulded or solid bars than for bars that have many ingredients. Imagine what would happen if the chocolate that surrounds the other ingredients was too runny. Too much would run off and the centre would not be properly coated. What if the chocolate pouring into the moulds was too viscous? The mould might not fill before the conveyor belt moved it along, or the chocolate might not flow throughout the entire mould, leaving air bubbles or gaps. Viscosity is a very important measurement in the production of chocolate bars.

Factors Affecting the Viscosity of Chocolate

An interesting aspect of Randy's job involves investigating how chocolate is ground down to the right smoothness. The pieces must be just the right size to ensure the chocolate product is smooth to taste. The size of pieces and the temperature of the chocolate affect its viscosity. When a lower viscosity is desired, scientists can add more fat to the chocolate. Fat, such as cocoa butter, coats the fine solid pieces in the chocolate so it flows more freely. Careful adjustments are made to obtain the right combination of smoothness, fat content, and temperature in liquid chocolate. This ensures its viscosity is perfect for each application.

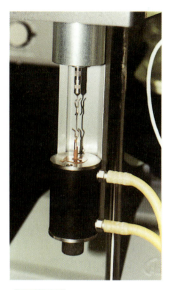

Figure 2

This specialized viscometer is measuring the viscosity of chocolate. A spindle rotates inside a sample of chocolate. If the chocolate has a high viscosity, there will be more resistance to the turning of the spindle. Viscosity tests are normally run on chocolate that has been heated to 40°C.

 **Hot Chocolate**

Look at how different the viscosity of chocolate can be for two different types of products. You will need chocolate from chocolate baking chips and from a moulded, solid milk-chocolate bar.

In separate glass measuring containers, carefully heat a small sample of each chocolate to 40°C. Measure the temperature of each sample to confirm they are the same.

🛑 Because chocolate burns easily, use a microwave oven at a medium setting to melt the chocolate.

- Stir each sample of liquid chocolate with a separate spoon.

1. Does one sample seem thicker and more viscous than the other? If so, which one?

- Fill each spoon with the liquid chocolate and hold it above the sample. Slowly pour the chocolate off the spoon onto a plate.

2. Which sample is slower to start pouring?

3. Which sample more quickly forms a pool of chocolate with a flat surface?

- Consider how these two types of chocolate are intended to be used.

4. How does this explain why their viscosity is different?

5. Why is it necessary to heat each sample to the same temperature before testing its viscosity?

Measuring Matter: Mass, Weight, and Volume

Using fluids—both liquids and gases—requires an understanding of their behaviour. You need to know how they behave when they are still, when they are moving, when something is moving in them, when they are pushed, or when they are pulled. Learning about these things requires the ability to measure matter.

Mass and Weight

"How much does it weigh?" "Let's check the weight of the candy." You hear expressions like these almost daily. Usually, when people use the term weight, they are referring to the measurement of mass. Mass and weight are not the same thing.

Mass is the amount of matter in an object and is used to measure many things, from food to mail. An object's mass stays constant everywhere in the universe. Mass is measured in grams (g), or units derived from grams, such as milligrams (mg) and kilograms (kg).

An object's **weight** is a measurement of the force of gravity pulling on the object. It is measured in newtons (N), named after Sir Isaac Newton. Because gravity is not the same everywhere in the universe, an object's weight varies according to where that object is in the universe. (See **Figure 1**.)

Because gravity is approximately the same everywhere on Earth's surface, people often use the words mass and weight interchangeably. Remember that mass and weight are different.

Volume

In addition to having mass and weight, matter occupies space. **Volume** is a measure of the amount of space occupied by matter. It is measured in cubic metres (m^3), litres (L), cubic centimetres (cm^3), or millilitres (mL).

Capacity is related to volume. It is a measure of the amount of space available inside something. People measure the volume or capacity of things such as fish aquariums, medical syringes, and ships' cargo holds.

Different types and quantities of matter are measured in different ways. Here are some techniques that you might use in your class.

Measuring Liquids

Liquids are measured by observing how much of a container they fill. A tall, narrow container (such as a graduated cylinder) gives the most accurate measurement. Look at the container from the side, with your eye level with the surface of the liquid. You might notice a slight curve at the edges of the surface where the liquid touches the

Figure 1

The downward pull (force of gravity) on an object on the surface of Earth is approximately 6 times as large as on the Moon. Because of this difference in gravity, objects on the Moon weigh 1/6 what they do on Earth. The *weight* of the object changes, but the mass is the same in both locations.

container. This "curved" surface is called the meniscus. Read the volume at the lowest place on the meniscus. Liquids are generally measured in litres (L) or millilitres (mL).

Measuring Volume of Solids — Rectangular Solids

Rectangular solids may be measured with a ruler, and their volume calculated using the formula

$$\text{volume} = \text{length} \times \text{width} \times \text{height}$$

Solids are usually measured in cubic metres (m^3) or cubic centimetres (cm^3), but may sometimes be given in litres (L) or millilitres (mL). Interestingly, 1 cm^3 is the same as 1 mL, so 1000 cm^3 equals 1 L.

Measuring Volume of Solids — Small Irregular Solids

The volume of a small irregular solid must be measured by **displacement**. In this technique, you choose a container (such as a graduated cylinder) that your small object will fit inside. Then pour water into the empty container until it is about half full. Record the volume

of water in the container, then carefully add the object. Record the volume of the water plus the object. Calculate the volume of the object using the formula:

$$\text{volume of object} = (\text{volume of water} + \text{object}) - (\text{volume of water})$$

Measuring Volume of Solids — Large Irregular Solids

To measure the volume of a large irregular solid, you will need an overflow can and a graduated cylinder (**Figure 2**). This measurement is best done over a sink. Fill the overflow can with water until water starts to run out of the spout. Wait until the water stops dripping, then place the graduated cylinder under the spout. Carefully lower the object into the water and observe what happens.

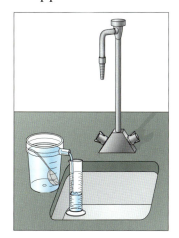

Figure 2

A volume of water equal to the volume of the solid will pour out of the spout and into the measuring cylinder.

Try This — Measuring Volume

Your teacher will provide you with several samples of matter and equipment to measure volume. Estimate the volume of each sample.

1. Record your estimates in a chart.
- Select the appropriate equipment for measuring the volume of one of the samples.
- Following the guidelines given above, find the volume of your sample.
- Share your results with the rest of your class.

2. Record the volumes of all the samples in your chart.

3. Which samples were you able to estimate quite accurately? Which were harder to estimate?

Understanding Concepts

1. Describe the relationship between mass and weight. Give an example of this relationship.

2. Imagine you have travelled to a planet that has twice the force of gravity of Earth. You have taken a solid with a mass of 1 kg with you. Describe its mass, weight, and volume on this planet, compared with that on Earth.

Design Challenge

What measurements will you need to make of the fluid in your Design Challenge?

SKILLS MENU
○ Questioning   ● Conducting   ● Analyzing
● Hypothesizing   ● Recording   ● Communicating
○ Planning

Relating Mass and Volume

How are volume and mass related to each other? If you double the volume of a substance, how will the mass change? Would the same volume of a different substance have the same mass? Which is heavier: a kilogram of feathers or a kilogram of lead? The answer seems obvious, but there is an important difference between feathers and lead. Equal masses of these substances have very different volumes.

Question

What does mass have to do with the amount of space (volume) a liquid occupies?

Hypothesis

1 Write a hypothesis about how you think the mass and volume of a **2C** liquid are related.

Experimental Design

In this investigation, you will measure volume and mass, plot them on the same graph, and draw conclusions about the relationship between them.

2 For each liquid, construct an observation chart with 3 columns, as in **Table 1**.

Materials

- apron
- safety goggles
- distilled water
- corn syrup
- saturated solution of salt water
- triple beam balance
- 150-mL or larger graduated cylinder
- small plastic pipette
- 150-mL beaker with a pour spout
- tissues

Procedure

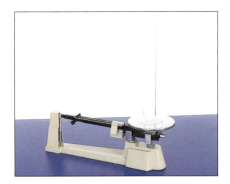

3 Measure the mass of an empty graduated cylinder.

✏ (a) Record this figure.

4 Obtain a sample of one of the three liquids from your teacher.
- Add 10 mL of the liquid to the graduated cylinder.

✏ (a) Record the mass of the cylinder and the liquid.

✏ (b) Calculate and record the mass of the liquid.

5 Continue to add the liquid, in 10-mL amounts, until you have a total of 100 mL in the cylinder.

(a) Calculate the mass of each new volume of liquid.

(b) Calculate the mass-to-volume ratio for the 20-mL and 60-mL volumes.

- Clean the cylinder.

Table 1

Volume of Liquid Added	Mass of Cylinder and Liquid	Mass of Additional Liquid
?	?	?
?	?	?

Figure 1

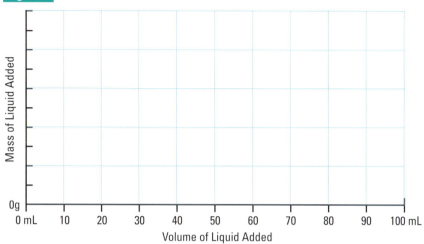

Making Connections

1. The line on your graph should go through the origin. Explain why.

Exploring

2. Make a prediction: add a line to your graph to show the relationship between mass and volume for copper. With your teacher's permission, repeat the investigation using copper pennies. You will have to find the volume of a copper penny. How well do your results match your prediction?

Reflecting

3. All water does not have the same composition. Why is it important to use distilled water in this investigation?

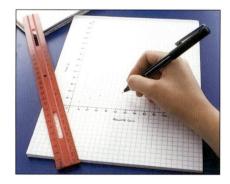

6 Make a line graph of your results. Put the volume of liquid added on the *x*-axis, and the mass of the liquid on the *y*-axis (**Figure 1**).

7C
• Draw the line of best fit through the points.

7 Repeat steps 4 and 5 with the other liquids.
• Add new data to the graph.
• Make a legend to distinguish the liquids.

Analysis

8 Analyze your results by answering the following questions.

(a) Why did you measure the mass of the graduated cylinder at the beginning of the experiment and not after the liquid was poured out?

(b) Calculate the mass of 1 mL of each liquid, by calculating the mass of 1 mL of liquid from each 20-mL amount that was added, then taking the average.

(c) How do the mass-to-volume ratios for the 20-mL and 60-mL volumes of each liquid compare to your answers to (b)?

(d) Your line graph illustrates the relationship between the mass and volume of three liquids. State this relationship in a way that answers the question at the beginning of this investigation.

Density: Another Property of Fluids

We have all seen the devastating effects of a spill from an oil tanker. (See **Figure 1**.) Cleaning up would be much more difficult if oil did not float on top of water. But why does oil float? We could say that oil is lighter than water, but what would that mean? A litre of oil is certainly not lighter than a glass of water.

Before we can compare fluids using the words "light or heavy," we must examine the same volume of each fluid. Thus, a litre of oil is lighter (has less mass) than a litre of water. When we compare the masses of the same volume of different substances, we are comparing their densities. Oil floats on water because it is less dense than water.

Calculating the Density of a Substance

Density is the mass of a substance per unit volume of that substance. It is expressed as grams per cubic centimetre (g/cm^3), kilograms per cubic metre (kg/m^3), or grams per millilitre (g/mL).

Density is calculated by dividing the mass of an amount of substance by its volume. The formula looks like this:

$$Density = \frac{Mass}{Volume} \quad or \quad D = \frac{M}{V}$$

For example, the cube of water in **Figure 2** has a volume of $1\ m^3$ and a mass of 1000 kg.

$$Density = \frac{1000\ kg}{1m^3} = 1000\ kg/m^3$$

Table 1	Densities of Common Substances	
	Density	
Fluids	**g/cm³ (g/mL)**	**kg/m³**
hydrogen	0.000 089	0.089
helium	0.000 179	0.179
air	0.001 29	1.29
oxygen	0.001 43	1.43
carbon dioxide	0.001 98	1.98
gasoline	0.69	
isopropanol (rubbing alcohol)	0.79	
vegetable oil	0.92	
distilled water	1.00	
seawater	1.03	
glycerol	1.26	
mercury (a metal)	13.55	
Solids		
wood (balsa)	0.12	
wood (pine)	0.5	
wood (birch)	0.66	
ice	0.92	
sugar	1.59	
salt	2.16	
aluminum	2.7	
limestone	3.2	
iron	7.87	
nickel	8.90	
silver	10.5	
lead	11.34	
gold	19.32	

Figure 2

One cubic metre ($1\ m^3$) of water is as heavy as a small car.

Figure 3

These balloons float because they are filled with helium.

Density Is a Property of Fluids and Solids

In Investigation 2.7, you calculated the density (mass-to-volume ratio) of water, corn syrup, and salt water. You found that this ratio was constant for each fluid. Each gas also has its own density. Helium gas will float on top of air (**Figure 3**), just as oil floats on top of water. This happens because helium is less dense than air.

Solids also have their own unique densities. Those that float in water have a density of less than 1.00 g/mL. Solids that sink in water have a density of more than 1.00 g/mL.

The densities of two substances can be used to predict which will float and which will sink.

Try This Build a Density Unit

This activity lets you take a closer look at the density of water.

• Design a cube that is 1cm² on each side. Leave the top of the cube open. Make it out of a light plastic, such as an overhead transparency. Be careful with your measurements. Accuracy is important. Fasten the cube together using cellophane tape. Fill the cube with water.

1. How much water will fit into this 1-cm³ container?

2. What is the mass of this amount of water?

3. Calculate the density of water.

4. Compare your calculated value with the value in **Table 1**. Explain any difference.

Understanding Concepts

1. Compare the densities of the three liquids in Investigation 2.7.

2. **(a)** What substance in **Table 1** is the most dense?

   (b) What substance is the least dense?

   (c) Give one use for each of these substances.

3. **(a)** List all of the solids that will float on water.

   (b) List all of the solids that will float on liquid mercury.

   (c) List all of the gases that will float on air.

Making Connections

4. Propane gas, which is used in many barbecues, is denser than air. It is also flammable. Propane appliances are often used in areas without electricity. Explain why a leak from a propane appliance is very dangerous.

Exploring

5. With your teacher's approval,
   (2E) design and conduct an investigation to find the density of a $1 Canadian coin from its mass and volume. Could density be used to identify counterfeit coins? Explain.

6. Make a 1000-cm³ container. What will be the dimensions of the length, width, and height? How much water will this container hold? What is the mass of this amount of water?

Design Challenge

Why is the density of water an important consideration when designing a submersible device or a boat lock?

Some Liquids Just Don't Mix

If you've made homemade salad dressing with vinegar and oil, you've likely noticed how one of these liquids tends to float on top of the other.

Question
Can different liquids float on top of each other?

Hypothesis

1 What will happen when corn syrup, vinegar, and cooking oil are **2C** placed in the same container? Predict whether a piece of cork and a plastic block will float or sink in the container. Write a hypothesis explaining your predictions.

Experimental Design
You will observe what happens when some common liquids are poured into the same container. You will also observe the relative densities of two solids: a cork and a plastic block.

2 Construct an observation sheet for this investigation, after reading the Procedure.

Materials
- apron
- safety goggles
- corn syrup
- white vinegar
- cooking oil
- piece of cork
- small plastic block
- 50-mL graduated cylinders, 2
- 15-mL measuring spoon
- paper towel
- triple beam balance
- food colouring

Procedure

3 Slowly pour 15 mL of any of the three liquids down the sides of one of the graduated cylinders.
- Rinse and dry the measuring spoon.

🖉 (a) Draw a diagram of your observations.

4 Repeat step 3 with the other liquids.

5 Using the second graduated cylinder and the triple beam balance, calculate the densities of vinegar, cooking oil, and corn syrup.

🖉 (a) Show your density calculations on your observation sheet.

🖉 (b) Write the density of each liquid beside its layer in the diagram.

6 Slide the cork and the plastic block into the column of liquid. Observe where they settle in the column.

🖉 (a) Draw and label their positions in the diagram.

Analysis

7 Analyze your results by answering the following questions.

(a) How do your observations of the liquids compare with your prediction?

(b) Were the results affected by the order in which you poured the liquids?

(c) How do the positions of the solids compare with your prediction?

(d) Write a brief explanation of the results you obtained referring to the properties of fluids.

(e) There are really four fluids in your density column. What is the fourth fluid?

(f) Compare the densities of the liquids to their position in the container. Are your results consistent with the density values? Explain.

Making Connections

1. What common food is made of the same three fluids, plus a little seasoning?

Exploring

2. Add 10 drops of 15% table cream to the column of liquids. Draw a diagram of your observations.

3. Estimate the densities of the cork and the plastic.

Reflecting

4. (a) Of the three liquids, which one took the longest to pour? Which poured most quickly?

(b) Propose a hypothesis **2C** relating two different properties of fluids.

5. Could food colouring added to the vinegar affect the density calculation of the vinegar? Explain.

Comparing Densities

You have already learned that every pure substance has its own characteristic density. Usually, solids have greater densities than liquids, and liquids have greater densities than gases. The particle theory can help us to understand this: the particles in solids are tightly packed together, held by the attractive forces between particles. There is relatively little space between the particles (**Figure 1**), so the substance tends to be dense. Liquids have a little more space between the particles, so are slightly less dense. Gases have very large spaces between the particles, so have the lowest density. When a solid is heated until it melts, it expands slightly and becomes less dense. So how can we say that each substance has its own characteristic density?

We need to understand that, unless we are told otherwise, the density given is for a substance in its most common state at room temperature. Copper, for example, is solid, and oxygen is a gas. There are a couple of exceptions to the "solids are most dense" rule. Mercury, a metal that is liquid at room temperature, is more dense than many solids and over 13 times as dense as water.

Water, also, is something of a special case. At some temperatures, liquid water is more dense than solid ice!

Using Pure Water as a Standard

Water is the most abundant liquid on Earth. We use the density of pure water as a standard by which the densities of all other fluids and solids are measured.

In the Try This activity in 2.8, you discovered (within experimental error) that the cubic centimetre you made holds 1 mL of water and that 1 mL of pure water has a mass of 1 g. The density of water is $1 g/mL$ or $1 g/cm^3$. This value is used to compare the densities of all other fluids and solids.

Table 1 in 2.8 lists the densities of some common substances. Comparing densities allows us to determine which substances will float on top of other substances. For example, carbon dioxide gas, a byproduct of respiration, is more dense than air. As a result, air floats on top of carbon dioxide, but both air and carbon dioxide sink below helium gas.

As you discovered in Investigation 2.9, more dense substances sink below less dense ones. Hydrogen is extremely explosive so firefighters need to know whether a hydrogen leak is likely to stay near the ground or float up, possibly getting trapped under a roof. Look at **Table 1** again and compare the density of hydrogen with that of air. Where do the firefighters need to worry about the escaped hydrogen?

Figure 1

This diagram shows how the particles of a substance gain energy and start to move as they are heated. We are looking at a fixed volume of the substance. You can see that the density and mass decrease as the temperature increases.

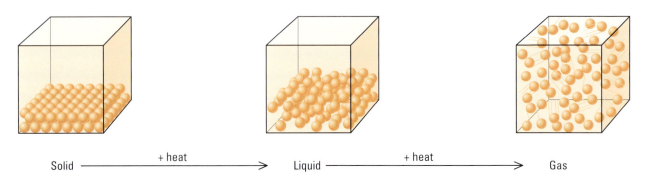

Solid   + heat   →   Liquid   + heat   →   Gas

What Portion of an Iceberg Is Submerged?

What happens when you add an ice cube to a glass of water? Not all the ice floats on the surface—some of the ice is below the surface (submerged). This also happens with icebergs. (See **Figure 2**.) By comparing the density of ice to the density of seawater, you can calculate how much of an iceberg is submerged:

$$\frac{\text{Density of ice}}{\text{Density of seawater}} \times 100 = \frac{0.92}{1.03} \times 100$$

$$= 89\%$$

Figure 3

Understanding Concepts

1. Make a general statement comparing the densities of solids, liquids, and gases.

Making Connections

2. Suppose alcohol, glycerol, water, and gasoline are placed in a tall container. Draw and label a diagram showing the order you would expect to find them.

3. Calculate the percentage of a piece of birch that will float above the surface of vegetable oil.

Exploring

4. **(a)** Great airships called dirigibles used to carry passengers between continents. The *Hindenburg* was the largest dirigible built and was filled with hydrogen gas. Research and report on what happened to the *Hindenburg*.

   (b) What were the advantages and disadvantages of using hydrogen gas in a dirigble?

Figure 4

Reflecting

5. Isn't ice supposed to float? Explain why the ice cubes in **Figure 4** have sunk to the bottom of the liquid.

The Ups and Downs of Buoyancy

What happens when you jump into water? The water pushes aside (displaces) to make room for you. (See **Figure 1.**) All fluids, gases as well as liquids, behave this way when an object is placed in them. A helium-filled balloon pushes aside air like a swimmer pushes aside water.

The fluid also pushes back in all directions on the object. The upward part of the force exerted by fluids is called **buoyancy**. Buoyancy is a property of all fluids.

Buoyancy is not the only force acting on an object in a fluid. The force of gravity (weight) also acts on an object. The effect of both forces operating together is described in the next section.

Figure 1
The volume of fluid displaced is equal to the volume of the object in the fluid.

Try This — Comparing Buoyancy and Gravity Forces

How are the buoyant force and the force of gravity related? What role do these forces play in getting an object to float?

- Weigh a lump of modelling clay (approximately 300 g), using the newton spring scale. You will need to tie a piece of string (about 0.5 m long) around the clay, leaving a loop to attach the scale. Record this weight as "weight in air."

Figure 2

- Fill a pail with tap water and lower the lump of modelling clay into the water. Submerge it completely but do not let it touch the bottom or sides of the pail. Record its weight when it is submerged in the water. Do not submerge the spring scale. (See **Figure 2.**) Record this as "weight in water."

1. What do you notice about the weight in air and the weight in water? What might the difference between these two values represent?

2. What is the buoyant force acting on the clay?

- Use this calculation and the words "force of gravity" to explain why the lump sank.
- Modify the shape of your lump of clay until it floats.
- When the clay floats, find the weight of the new shape in air.

3. What do you notice about this weight?

- Let this new clay shape float.

4. Would you be able to find the weight in water now? Explain.

- Add marbles, one at a time, until the clay shape is one marble away from sinking. Record the total mass of marbles that your shape will hold.

5. What design similarities exist among all of the class's floating clay structures?

6. Does each floating clay shape hold the same mass of marbles? How does the shape that holds the most marbles compare to the shape that holds the least?

Archimedes' Principle

About 250 B.C., the king of Syracuse, on the island of Sicily, suspected that his goldsmith had secretly kept some of the gold meant for the royal crown and replaced it with a cheaper metal. The king asked Archimedes, a Greek mathematician, to determine whether the crown was made of pure gold.

Here's how Archimedes solved the problem. He found that the crown appeared to weigh less in water than a bar of pure gold with the same mass. This meant there was a greater buoyant force on the crown. Archimedes realized that the crown displaced more water than the gold bar. (See **Figure 3**.) Since the volume of each object was equal to the volume of water it displaced, the volume of the crown was greater than that of the gold bar. Therefore the crown had a lower density and was not made of pure gold.

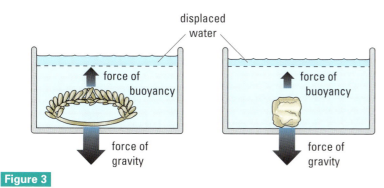

Figure 3

The buoyant force equals the weight of the fluid that the immersed object displaces.

The key idea that Archimedes realized is still known as **Archimedes' principle**: The buoyant force on an object immersed in a fluid is equal to the weight of the fluid that the object displaces. According to legend, Archimedes thought of the idea while taking a bath. He was so happy that he leapt up and ran through the streets crying "Eureka!" (which means "I have found it!").

Understanding Concepts

1. Why did the king's goldsmith mix a less dense material with the gold from the crown?

Making Connections

2. **(a)** How would the buoyant force acting on the floating clay shape change if it were immersed in seawater?

   (b) How would this affect the amount of weight it could carry?

Exploring

3. Measure the amount of water that was displaced with the clay lump and the floating clay shape. How do these amounts compare?

Reflecting

4. How can you modify a dense solid substance to make it float in a less dense fluid?

5. Think back to Investigation 2.9. Explain the behaviour of the cork and plastic block, using the terms "buoyant force" and "density."

Design Challenge

How might your knowledge of buoyancy assist you in the Challenge of designing a submersible device or a boat lock? Explain.

Figure 4

Logs are buoyant: they float on water.

How and Why Do Things Float?

Figure 1

Hundreds of 2-L plastic bottles are given a second life as part of a dock flotation system. The sealed bottles are stacked in the float drum (inset) and add volume without much weight. Reusing the bottles reduces waste in landfills.

Remember the Try This activity in 2.11? You took a dense material (clay) and made it float. Shipbuilders do this all the time. They take steel, which has a density eight times that of water, and make it into a floating boat. Just as you changed the shape of the clay, ship engineers design the hull of a steel ship to contain a large volume of air. The overall density (total mass divided by total volume) of the whole ship, including the hollow hull, is less than the density of water. Like the pop-bottle floating dock in **Figure 1**, the ship is buoyant. It floats.

Figure 2

Forces Acting on a Floating Object

If the upward buoyant force on an immersed object is greater than the downward force of gravity (weight of the object), the object will rise. If the buoyant force is less than the object's weight, it will sink. If the two forces are equal, the object will not move up or down. (See **Figure 2**.)

But Archimedes' principle says that the buoyant force on the object equals the weight of the fluid it pushes aside. So the object will rise or sink depending on whether it weighs less or more than the fluid it displaces. Since they have equal volumes, the object will rise or sink depending on whether it is less or more dense than the displaced fluid.

It is interesting to note that buoyancy depends on gravity, because buoyancy is a result of the weights of various substances. Without gravity there would be no buoyancy. We say that the object has **positive buoyancy**, **negative buoyancy**, or **neutral buoyancy** according to whether it rises, sinks, or remains level in the fluid. (See **Figure 3**.)

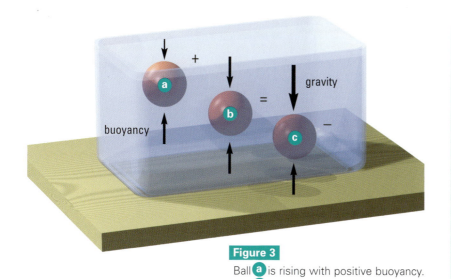

Figure 3

Ball **a** is rising with positive buoyancy. Ball **b** is stationary with neutral buoyancy. Ball **c** is sinking with negative buoyancy.

Buoyancy in Air

The buoyant force acts on objects immersed in a gas the same way it acts on objects immersed in a liquid. The densities of gases and liquids are very different. Air is almost 800 times less dense than water. That is why you would need an enormous helium balloon to rise through the air, but only a small life jacket to float on top of the water. To support your weight, the balloon must displace a much greater volume of less dense fluid.

Safe Floating Levels

The load lines on a ship are called Plimsoll lines. These lines, or numbers, show a safe floating level when the ship is fully loaded. (See **Figure 4**.) They are named after Samuel Plimsoll, a British politician. Around 1870, he helped develop a law that every British ship should have these load lines. Before this law, many owners overloaded their ships, and many ships sank. By the end of the 1800s, every ship in the world was using Plimsoll lines.

Understanding Concepts

1. If you were to do the Try This in 2.11 again, using cooking oil instead of tap water, how would the buoyant force change?

2. Explain, using scientific terms, why overloading a ship might cause it to sink.

Making Connections

3. Give examples of three real-life situations that match the diagrams in **Figure 3**.

Figure 5

4. Why do floating candles (**Figure 5**) float higher in the water as they burn?

Exploring

5. A hard-boiled egg in water is negatively buoyant—it sinks. Using salt, alter the buoyant force of tap water until the egg becomes positively buoyant (floats). How much salt do you use?

● Design Challenge

How can you apply this additional knowledge of buoyancy to enable a submersible device to be positively or negatively buoyant?

Figure 4

Plimsoll lines on the hull of a ship indicate the depth to which it may be legally loaded.

Another Way to Measure the Density of a Liquid

In Investigation 2.9, you compared fluid densities without making precise measurements. An instrument that measures the density of liquids is called a **hydrometer**. It is designed to float at different heights in liquids of different densities. The upper part of the hydrometer is marked with a scale. Most hydrometers are long and thin, with a weight at the bottom end.

Problem
In the production of maple syrup (**Figure 1**), there isn't enough time to measure the mass of a volume of boiling sap and then calculate its density. A faster method is needed.

Design Brief
Design and build a hydrometer. Determine the effectiveness of your homemade hydrometer by comparing it to a commercial model.

Design Criteria
Design a hydrometer that floats at different heights in different liquids, indicating the density of each liquid. The instrument must float upright. Your design should use only the materials made available to you.
• The instrument must be calibrated (a readable, linear scale must be present on the instrument), using water as the standard.

Materials
• apron
• safety goggles
• wooden dowel 17 mm in diameter
• fine, permanent marker
• bucket filled with water
• short, galvanized screw
• galvanized nut
• screwdriver
• centimetre ruler
• drill with 6-mm drill bit
• mitre box
• junior hacksaw
• saturated salt solution
• water
• pickling vinegar
• commercial hydrometer
• tall, thin cylinder

Build

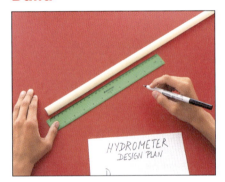

1 Design your hydrometer.
• After your teacher has approved your design, build your hydrometer.

Test

2 Test your hydrometer and modify it as necessary.

✎ (a) Record your modifications.

3 Use the commercial hydrometer to measure densities of the liquids above.
• Wipe off the hydrometer between liquids.

✎ (a) Record your observations.

Figure 1
The sap used to produce maple syrup is cooked in huge evaporating pans.

 Use tools and materials properly. If you are not certain how to use equipment safely, ask your teacher.

4 Calibrate your hydrometer by floating it in the three liquids.
• Mark the height for each, and write the densities on the instrument.

   (a) Draw and label your completed hydrometer.

Evaluate

5 Evaluate your hydrometer by assessing how well it met the design criteria.

   (a) Did your hydrometer float at different heights in different liquids?

   (b) Did you use only the materials made available to you?

   (c) Is your hydrometer calibrated to indicate the density of liquids?

   (d) If you had to modify your hydrometer, why did you have to make those modifications?

Analyzing

1. When will a hydrometer float higher? When will it become more immersed?

2. What limitations might exist with your homemade instrument because it is constructed out of wood?

3. Rank the liquids by density, from least to greatest.

4. How effective is your hydrometer in measuring liquids with similar densities?

Making Connections

5. How could a hydrometer be used during the manufacture of maple syrup?

6. What happens to sap that is boiled too long?

Exploring

7. Make sugar solutions with concentrations of 200 g/L, 300 g/L, and 400 g/L. Measure the density of each solution with the commercial hydrometer. How does measured density relate to the mass of sugar dissolved in each solution?

Reflecting

8. Explain the meaning of the following hydrometer readings: 1.03, 0.87, and 1.00.

Design Challenge

Is density an important measurement to consider when designing your Challenge? Explain.

From Bladders to Ballast: Altering Buoyancy

Fish and some aquatic plants (**Figure 1**) have adaptations that alter their buoyancy in water. Design features of ships, submarines, hot air balloons, and scuba equipment also alter their buoyancy, allowing them to move vertically in water or air. Natural and engineered methods of altering buoyancy have many similarities.

Nature's Method

Fish control their depth in water using swim bladders containing air. (See **Figure 2**.) They can get more gas into their swim bladders either by gulping air in at the surface of the water or by releasing dissolved gases from their blood.

(a) By expanding and contracting their swim bladders, fish can change their level in water. How does this enable fish to become more or less buoyant?

(b) Speculate why fish need to descend or rise to different water levels.

(c) How would an adaptation such as air bladders benefit seaweed?

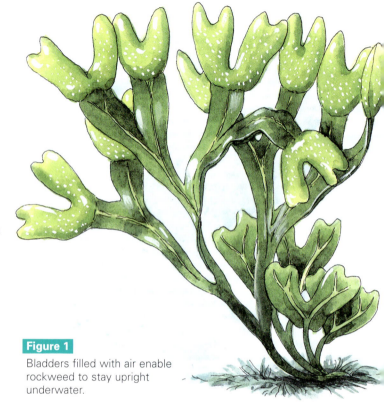

Figure 1
Bladders filled with air enable rockweed to stay upright underwater.

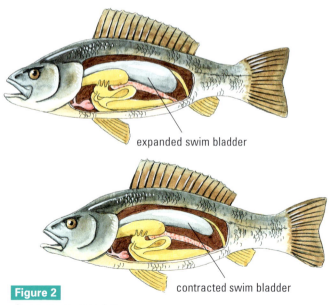

expanded swim bladder

contracted swim bladder

Figure 2
Swim bladders in fish

The Human Method

The human body has an overall density very close to that of water. When immersed in water, a relaxed swimmer with filled lungs is positively buoyant. Wearing a wet suit further increases buoyancy. Divers wear weight belts to give them more density and less buoyancy. (See **Figure 3**.)

A scuba diver might also use a buoyancy compensator vest to alter her buoyancy in the water. If she wants to sink down, she releases air from the vest, thus making herself more dense. The vest also enables the diver to stop descending and become neutrally buoyant. To swim back to the surface, the diver can blow air into the vest to decrease her density and increase her buoyancy.

(d) In what ways are fish bladders and buoyancy compensator vests similar?

Controlling Ballast

Ballast is any material carried on ships, submarines, hot air balloons, or dirigibles (air ships) that acts as weight and alters buoyancy. The ballast helps the vessel to be stable and to travel at the appropriate level in the fluid. Tanks of water often provide the ballast for ships and submarines. In a hot air balloon or dirigible, ballast may be sand or water.

A fully loaded ship floats lower in the water and is more stable than an empty one. When a ship unloads its cargo, water is taken in as ballast to maintain stability in the water. When the ship takes on a new cargo, it pumps out the water it was using as ballast.

Hot air balloons and dirigibles are immersed in air. They use ballast to control their buoyancy. The weight of the fuel and passengers is calculated before deciding how much ballast is needed. Once in flight, the crew can increase buoyancy in one of two ways. It can increase the volume of gas in the balloon part by heating the air or adding more helium. Alternatively, it could release some of the ballast. To decrease buoyancy, the crew either has to reduce the volume of gas (by letting it cool or releasing some of it) or pick up more ballast (perhaps by scooping water from a lake in a bucket on a long rope).

If the ballast tanks are filled with water, a submarine will descend. (See **Figure 4**.) When the submarine reaches the desired level, some water in the ballast tanks is pumped out, to be replaced with air. This continues until the submarine stops sinking and becomes neutrally buoyant. To make the submarine surface, or float at a higher level, more water in the ballast tanks is replaced with air.

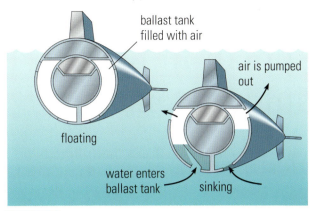

Figure 4

Changing the amount of water in its ballast tanks makes a submarine sink or rise.

(e) Do ships in the Great Lakes need more or less ballast than ocean-going ships? Give reasons for your answer.

(f) What could the crew of a hot air balloon do to raise the balloon above some trees?

(g) What factors affect how much water is carried in a submarine's ballast tanks?

(h) Where might the air come from to replace the water that is pumped out of a submarine's ballast tanks?

Making Connections

1. If the force of gravity (weight) on a scuba diver is 600 N, what should the buoyant force be if the diver wants to

   (a) descend?

   (b) rise to the surface?

2. You are asked to add ballast to a helium-filled balloon until it will float in the centre of the room.

   (a) What could you use as ballast?

   (b) What happens if you add or remove some of this ballast?

Figure 3

Design Challenge

If you want a device to move up or down in water or air, the buoyant force needs to be altered. How might the information on these two pages assist you in designing your Challenge?

Human Impact on Natural Fluid Systems

Earth is surrounded by fluids, both liquids and gases. These fluids are the water we drink and the air we breathe. Sustaining the health of these systems is essential. Healthy fluid systems are necessary for the survival of natural ecosystems as well as for human health. At the same time, many economic activities (fishing, shipping, and tourism) depend on Earth's fluid systems. We must manage Earth's fluid systems in a way that benefits the environment, the economy, and society. Despite their importance, these fluid systems have been abused and neglected.

Ballast Beware

Each year, many different cargos from around the world arrive by ship in Canadian harbours. These ships carry almost anything you can imagine, from raw materials such as grain and oil to electronic products and automobiles. In each port where they unload cargo, these ships take in water for ballast. When they load cargo in the next port, they pump out the water.

The problem is that the dumped ballast contains more than just water. It can contain living things from other parts of the world that are not normally found in Canadian waters. These are referred to as exotic species. Exotic species include species of plants (especially algae), fish, and microscopic organisms. (See **Figure 1**.)

When conditions are favourable, exotic species can multiply rapidly in their new environment. Usually there are no natural predators, so growth can go unchecked for years—until a predator develops or a method of control is found.

Exotic species of fish brought to the Great Lakes in ballast water include the ruff and the goby. They compete with native populations of fish for food and shelter, upsetting the balance of the ecosystem. This is a concern because fish are an important source of food, recreation, and income for Canadians.

Ballast water may also contain oil or other pollutants. After some oil tankers empty their shipments of crude oil, they take on seawater as ballast in the same tanks that held the oil. When the tankers are filled again, this contaminated ballast is pumped into the local waters.

Zebra Mussels in Canadian Waters

In 1985 or 1986, ballast water from a European freighter accidentally brought zebra mussels to the Great Lakes. They were first discovered in Lake St. Clair and Lake Erie and have now spread to inland lakes and rivers throughout Ontario and Quebec. The effects of zebra mussels have been widespread. They reduce the food supply available to fish by feeding on plankton. They clog water intake pipes and screens of factories and water treatment plants, affecting the water flow rate. (See **Figure 2**.) If less water enters the intake pipes, not enough water is available for use. Because zebra mussels are filter feeders, they contain high concentrations of toxins. Any animal that feeds on them will ingest these toxins, which are then passed up the food chain. Millions of dollars are spent annually on zebra mussel control.

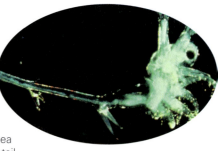

Figure 1

The spiny water flea, a tiny crustacean, arrived in Canada in ballast water around 1993. It feeds on plankton, an important food source for native small fish. Many native fish cannot eat the spiny water flea because of the spines on its tail.

Presentation Ballast Water Management ⑧ⓓ

Statement

Action should be taken to stop the pollution of Canadian waterways by contaminated ballast water.

Sample Opinions

From a utilities official

Ballast water pollution has introduced zebra mussels into our lakes and rivers. This water is used for electricity generation and for drinking water. The zebra mussels are blocking the water intake pipes. It's a disaster for utility companies. The technology and labour needed to remove the mussels from our intake pipes is very expensive. This cost is passed on to everyone who uses electricity or water. Who knows what further damage may be caused by the introduction of other exotic species?

From a port official

Ships are designed to float at a certain depth, so they have to carry ballast when they have no cargo and empty the ballast when they load up. Shipping is a very competitive industry. If Canadian ports make too great a demand on shipping companies to control their ballast water, they'll go to other ports. Thousands of Canadian jobs could be lost, and the cost of shipping to and from Canada will increase. Canadian consumers and exporters will have to pay more.

From the captain of a Great Lakes fishing boat

The damage has already been done. Many species of fish have already been almost wiped out: out-competed or eaten by exotic species from ballast tanks. If there's to be any hope of the fish stocks recovering, we must stop exotic species from getting into our waterways. But ships have to have ballast. Maybe there's some way of cleaning the ballast water.

From a cottager

The water in our lake is much cleaner than it used to be. That may be because of the zebra mussels: they're all over our dock's piers. But we can't catch the same kinds of fish that used to be in the lake. I think they're been eaten by the introduced species. I prefer the way the lake used to be. Let's stop ballast pollution.

What Do You Think?

• Read the sample opinions and evaluate each one. Record the main points under appropriate headings in a chart. Add other points of your own under these headings.
• Design a solution to the problem and prepare a proposal outlining your solution.
• Present your proposal to the class. In your presentation, explain how your solution would reduce the human impact on the environment.

How Does Temperature Affect Viscosity and Density?

Have you noticed that it's easier to run after a good warm-up? Fluids run more easily when they're warm, too. Viscosity, density, and buoyancy all change with changes in temperature.

First, let's discuss fluids other than water. Have you heard the expression "slower than molasses in January"? This describes the increase in resistance to flow that fluids experience when the temperature drops. As heat is taken away from a fluid, its particles slow down and come closer together. This causes the fluid to contract—its volume decreases. Thus the fluid's density will increase. (Remember, $D = M/V$. Since M stays the same and V gets smaller, then M/V will get bigger.) Viscosity will also be affected. When the particles slow down and come closer together, the forces of attraction between them will increase and so make it harder for them to flow past each other. Thus, viscosity increases at lower temperatures.

As you would expect, the opposite occurs when the temperature rises. When heat is added to a fluid, its density and resistance to flow decrease.

The reaction of air to temperature change explains the behaviour of hot air balloons. As air is heated and released inside the balloon, the balloon rises. This happens because hot air is less dense than the surrounding air, so it rises to float above the cooler air. As the air inside the balloon cools, it becomes denser, and the balloon descends. Periodic bursts of heat keep the balloon aloft.

Water: A Special Case

Water behaves differently from other fluids when the temperature changes. You may have noticed during a dive into a lake that the top layer of water feels warmer than the lower layers. During the summer, warmer water, because it is less dense, floats on top of cooler water. But as the temperature of water drops below 4°C, water becomes less dense again! **Table 1** illustrates this.

Table 1	
Temperature of Pure Water	**Density (g/cm³)**
100°C	0.958
20°C (room temperature)	0.998
4°C	1.000
0°C	0.92

Ice floats because it is less dense than liquid water. Water is most dense at 4°C.

This unique property of water keeps lakes from freezing solid in the winter. As the water cools, it sinks to the bottom. The deepest part of the lake will be at 4°C: a liquid. This enables aquatic life to survive. The ice on top of a lake insulates the water beneath. (See **Figure 1**.) Only shallow ponds freeze solid in the winter.

The viscosity of water also changes with temperature. Water at 0°C is approximately seven times more viscous than water at 100°C.

Figure 1

Water becomes lighter as it freezes. At 4°C, it is most dense and falls below cooler, frozen water.

Water at 0°C

Water below 4°C

Explaining the Effect of Temperature Changes Using the Particle Theory

In a solid, particles are closely packed together and held in a rigid structure by forces of attraction between them. The particles can move, but only by vibrating in the same place. When a solid is heated, the particles gain more energy and vibrate faster. As more heat is added, this speed of vibration becomes so fast that the force of attraction cannot hold the particles together. The rigid structure of the solid falls apart, melting occurs, and a liquid is formed. In a liquid, the particles are slightly less tightly packed together (less dense) than in a solid.

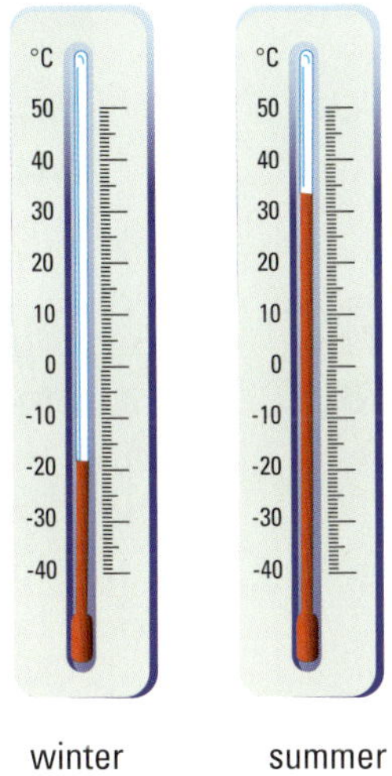

Figure 2

As the temperature of a liquid increases, the spaces between its particles become greater, and the volume of the liquid increases. When placed in a glass tube, alcohol increases or decreases in volume as the temperature fluctuates. That's how thermometers work.

winter   summer

As more heat is added to a liquid, the particles move even faster. The forces of attraction between them are broken, and the particles are able to move in all directions, leaving larger spaces in between. The particles take up more space or volume (**Figure 2**), making the density lower. Particles eventually escape from the liquid, and a gas is formed. (See **Figure 3**.)

The reverse process occurs when heat is taken away from a gas or a liquid as its temperature decreases.

As the density of a fluid decreases with a rise in temperature, so does the force of buoyancy that the fluid exerts on an immersed object. Why is that? Because the displaced fluid weighs less at a higher temperature. The viscosity of the fluid also decreases as the attraction between its molecules weakens.

Table 2

	Volume	Density	Viscosity	Buoyancy
Temperature ↑	?	↓	?	?
Temperature ↓	?	?	?	?

Figure 3

Firefighters often use water in the form of a fine spray. This fine spray absorbs heat from the fire faster than a solid stream of water would. As heat is absorbed, steam is produced. Steam occupies a larger volume and displaces the air that is fuelling the fire.

Understanding Concepts

1. Copy **Table 2** and complete it by adding up or down arrows to indicate how each property changes with temperature.

2. Use the words *mass*, *volume*, and *density* to distinguish between gases, liquids, and solids in terms of the particle theory of matter.

3. Use the particle theory to explain the effects of temperature changes on the cooking oil in Investigation 2.4.

4. How does water behave differently than other fluids when the temperature changes?

Making Connections

5. In many aircraft, oxygen masks are stored in compartments above the passengers. The oxygen for these masks is stored as a liquid. When it is needed, it is warmed up until it is a gas. Explain why oxygen is stored as a liquid rather than a gas in this situation.

6. Will ships float lower or higher in tropical waters? Explain your answer using the words *buoyancy* and *density*.

7. Suggest two examples of a substance changing its temperature in the natural world. What happens to its viscosity and density in each case?

Exploring

8. Lava, brought to the surface when a volcano erupts, changes in viscosity as it cools. Research how lava is produced. Explain the viscosity changes it undergoes.

Design Challenge

How will temperature changes affect the fluid in your Challenge?

Fluids and the Confederation Bridge

Imagine the challenge of building a structure over 12 km long, across a storm-tossed stretch of ocean. The structure has to last 100 years and be safe for motorists to drive on. That was the task facing the engineers on the Confederation Bridge project. The connection from Prince Edward Island to New Brunswick opened on May 31, 1997. The 12.9-km bridge crosses the Northumberland Strait and is the world's longest bridge to cross ice-covered waters.

Some of the challenges facing the bridge designers are described below. In order to solve these challenges, engineers required a knowledge and understanding of the properties of fluids and how forces and motion affect fluids.

Barges

During the building of the bridge, much construction took place from rectangular floating vessels called barges. Many activities were carried out from the barges, including positioning the pier bases and cementing them to the bedrock, and transporting

Figure 1

The *Svanen*, a barge with a floating crane, was used to carry and install the bridge sections.

supplies to workers. One barge was even equipped with a helicopter landing pad.

So building of the bridge could continue during the long winter season, sections for the bridge had to be first built on land and then floated out to their final position. Each bridge section consisted of a pier and girders and weighed about 7500 t.

(a) What forces must an engineer consider when designing a barge?

(b) Why is it crucial that barges float and be stable during all of the activities that are carried out from them?

(c) What could the engineers do to ensure a barge was stable before use?

Water and Ice

Water constantly exerts force on the bridge piers. Some days enormous waves crash into the piers. This pushing force increases as the water freezes and ice slams into the piers.

The ice in the Northumberland Strait was a major concern for the engineers designing the Confederation Bridge piers. A model of this situation was constructed. In an enormous basin, several centimetres of ice were produced. A model of a pier attached to a bridge was pushed through the ice and across the basin. The speed at which the pier was pushed was carefully controlled to mimic actual water current conditions. Engineers recorded the investigation on videotape and took measurements throughout the testing. The

Figure 2

Exploring

1. Research the design and building of the Confederation Bridge.

   (a) What structure was added to the base of each pier to break up ice sheets drifting down the Northumberland Strait during the winter?

   (b) How was the *Svanen* brought to Canada? What are some interesting facts about this amazing vessel?

results were used to determine the forces the piers must be able to withstand.

(d) Is the force of the water on the piers the same at the water surface as it is 30 m below the surface?

(e) Why did engineers construct a model of the piers?

Winds

High winds posed another challenge to the bridge designers. They considered how air would flow around the bridge and how winds would affect the bridge itself. How the winds would affect the vehicles using the bridge was also a major concern, and barrier walls on each side of the roadway were designed to minimize this effect.

(f) Suggest some design features that might reduce the amount of air turbulence around the bridge.

Concrete

The concrete used in the design of the bridge also concerned the engineers. To make a pier that could withstand collisions from ice and possibly ships, a special high-strength, low-water concrete was used. The concrete had to be pumped through pipes and poured into forms to make the pier shapes. The engineers changed the viscosity of the concrete by adding special products. This allowed the concrete to remain liquid longer.

(g) Why did engineers add special products to the concrete used in the bridge?

(h) Why was it necessary that the concrete fill the entire form it was poured into?

Figure 4

An apparatus that looks like the Canadarm was used to pour concrete.

Figure 3

2.18 Inquiry Investigation

SKILLS MENU
- Questioning
- Hypothesizing
- Planning
- Conducting
- Recording
- Analyzing
- Communicating

How Fluids Handle Pressure

"I'm under so much pressure!" How often have you heard that phrase? An upcoming test or too much to do in a short period of time can make people say they're under a lot of pressure. Fluids can be under a different sort of pressure. What happens to fluids under pressure? What effects can we observe?

Question

2B **1** After reading through this investigation, write a question that you will be trying to answer.

Hypothesis

2C **2** Write a hypothesis for this investigation.

Experimental Design

In this investigation, you will investigate the effects of exerting pressure on air and water in closed systems.

3 Copy **Table 1** and complete it.

Materials
- apron
- safety goggles
- 20-mL syringes, 2
- 5-mL syringes, 2
- 3-cm lengths of 6-mm tubing, 3
- 40-cm length of 6-mm tubing
- straight connector
- T-connector

Use equipment only as instructed. Be careful when working with syringes under pressure.

Procedure

4 Connect both 20-mL syringes with a 3-cm piece of tubing.

(a) Can you pull one plunger back? If not, what do you have to do to one plunger before connecting the tubing?

(b) As you depress one plunger, what happens to the other one?

(c) Try moving one plunger and holding the other one still. What happens?

(d) What fluid are you investigating here?

5 Repeat step 4 using the 40-cm piece of tubing.

(a) Do you notice anything different happening when you move one plunger?

6 Use a straight connector and two short pieces of tubing to join both 20-mL syringes.

(a) What is different about the movement of the plungers compared to the setup in step 4?

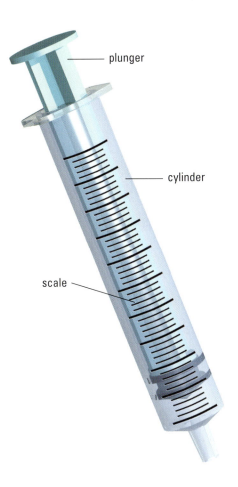

plunger

cylinder

scale

Table 1	Investigating Water and Air Pressure		
Investigation	**Setup Used**		**What Happened?**
1. Air pressure	(a) large syringe + 3 cm tubing + large syringe		?
	(b) large syringe + 40 cm tubing + large syringe		?
2. Water pressure	(a)	?	?
	(b)	?	?

7 Join one 20-mL syringe to both 5-mL syringes using the T-connector and three short pieces of tubing.

(a) Record the volume of air that starts in the large syringe.

(b) As you depress the plunger of the large syringe, what happens?

(c) Write your prediction for what you think would happen if you were to use water in steps 4 to 7.

8 Repeat steps 4 to 7 with the system full of water. You must ensure that there are no bubbles present.

Analysis

9 Analyze your results by answering the following questions.

(a) What would happen if the tubing on the syringes, or the plunger in the syringe, did not make a tight seal?

(b) What differences did you observe between the two different fluids when you applied pressure to them?

(c) Write a report explaining your results.

Making Connections

1. Design a syringe setup to **3D** raise an object attached to one plunger. Draw a sketch of your design.

2. Needles are attached to syringes when injections are given to people or animals. How do health care personnel get air out of the syringe?

Exploring

3. Use what you have learned about syringes to explain the benefit of having two lungs instead of one.

🟡 Design Challenge

How has this investigation on pressure helped you in designing your Challenge?

Confined Fluids Under Pressure

What are confined fluids? They are any fluids in a closed system. Confined fluids can move around within the system, but they cannot enter or leave the system. The blood moving through your body is a confined fluid (providing you don't cut yourself!) and so is the air in an air mattress (**Figure 1**). When fluids are confined, they have some very interesting effects.

Figure 1

In Investigation 2.18, you discovered that moving one syringe causes another syringe to move. In other words, applying a force to one part of a fluid system results in movement in another part of the system. The force was transmitted through the fluid to another moveable part, some distance away. This is one effect of a pressurized fluid system: forces can be applied in one place and have an effect somewhere else—even in another direction. The brakes in a car are an example of this. The driver presses down on the brake pedal, which pushes fluid through the brake lines toward the wheels, where the brake pads are forced against the moving wheels (**Figure 2**).

You might have noticed a difference in the effects of water and air in Investigation 2.18. Did you notice that there was a short delay, or bounce, in the air-filled system, whereas the water-filled system reacted immediately? Why might this be? Can you explain it using the particle theory? Think of the particles in liquids and gases, and the spaces between them.

Pushing the brake pedal forces a piston against the hydraulic brake fluid in the main cylinder.

The brake fluid is forced out of the cylinder into the brake lines toward the wheels.

The brake pads are forced against the moving wheels.

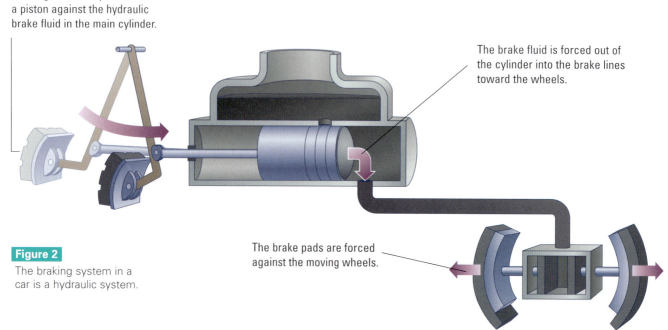

Figure 2

The braking system in a car is a hydraulic system.

Using the Particle Theory

We can use the particle theory of matter to understand what happens to confined fluids when an external force is applied to them. Remember that in a liquid, the spaces between the particles are already very small. When an external force is applied, only a small decrease takes place in the liquid's volume.

In a gas, the particles are far apart from each other. In order for the force to be transmitted from one particle to another, the volume the gas occupies must be reduced. This is referred to as compression. When an external force is applied to a gas, the force will push the particles closer together. This is why there is a delay in the air-filled system. It takes time to compress the air. Gases are very **compressible**. The change in volume of a liquid under pressure is so small that liquids are only very slightly compressible. We can say that liquids are almost incompressible.

There is another effect that can occur when a force is applied to a gas or a liquid. Its state can be changed. By increasing the pressure on a gas, the particles can be pushed close enough together that the gas will change to a liquid. For example, propane is normally a gas, but in a barbecue tank, under pressure, it is a liquid (**Figure 3**). Similarly, a liquid may be compressed until it changes into a solid.

Figure 3

Barbecue tanks contain liquid propane. Putting propane under pressure and storing it as a liquid allows propane tanks to hold more.

Understanding Concepts

1. Using the particle theory and this new information on pressure, explain the results you obtained with syringes in Investigation 2.18.

2. What evidence from the syringe investigation supports the particle theory?

3. The first paragraph on these pages mentions the "interesting effects" of fluids under pressure. What are these effects?

Making Connections

4. Compare liquids and gases in terms of their compressibility. 6C Draw a diagram to help your comparison.

5. Cars use a liquid brake fluid to transmit a force from the brake pedal to the brake pads. If air were used instead of a liquid, how different might pushing on the brake pedal feel? Explain.

Exploring

6. The whale is a mammal that has 4A adapted to aquatic life. Some whales dive to depths greater than 2000 m, deeper than most submarines can dive. Research how the respiratory system of a whale allows it to perform deep dives despite the enormous pressure of the water.

Reflecting

7. Why would you want to put a fluid under pressure?

Design Challenge

Describe how you will put a fluid under pressure in your Design Challenge.

Pressurized Fluid Systems: Hydraulics

Hydraulics is the word we give to confined, pressurized systems that use moving liquids. Hydraulic systems use liquids under pressure to move many things. Huge amounts of soil at a construction site can be moved with hydraulic machinery such as backhoes and excavators.

What Makes Up a Hydraulic System?

The liquid put into a hydraulic system is called the **hydraulic fluid**. The hydraulic fluid in the system in **Figure 1** is oil. Oil from the tank is sent along a conductor (a hose, tube, or pipe) to a pump where it is pushed into a **cylinder**. The cylinder resembles a giant syringe. The oil pushes up the **piston** in the cylinder like a plunger moving inside a syringe. This upward movement of the piston can be used to do work by moving something else.

A valve placed between the pump and the cylinder controls the flow of oil. This allows the piston to be moved a little or a lot. A second valve, placed between the cylinder and the tank, can be opened to allow the oil to flow back into the storage tank, pushed by the weight of the piston as it moves back down. The fluid is circulated through the system and is not used up.

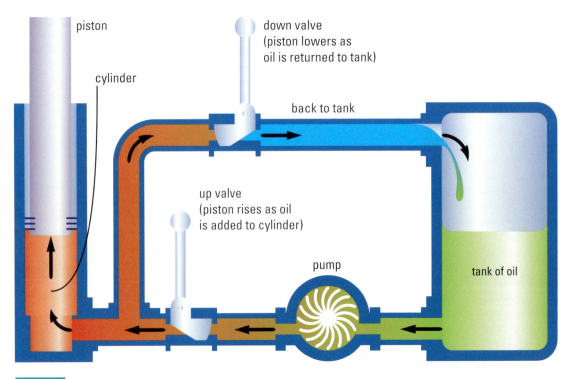

Figure 1

This hydraulic system could be used to raise and lower a snowplow blade or a car hoist.

Pumps and Valves

A pump is used to create a flow of fluid. Pumps often make fluids flow against gravity. They are found in car engines (**Figure 2**), gasoline pumps at gas stations, dishwashers, and many other machines.

Valves control the flow of fluid. There are many different types, but they all have a similar function: to keep a fluid flowing in the desired direction. When you turn on a water tap, you are opening up a valve. There are numerous places where valves are found, including tires, soccer balls, and the human heart. Can you think of any others?

The Heart: A Pressurized Fluid System

Your heart is also a pump. (See **Figure 3.**) It beats over 100 000 times a day to push blood through the veins and arteries that make up your circulatory system. There are four chambers in the heart: right atrium, right ventricle, left atrium, and left ventricle. The chambers in the upper part of the heart are separated from those below by valves. The valves allow blood to flow in only one direction. Knowledge of how fluid flows flow through valves was used to design artificial heart valves.

Blood pressure ensures that all of your organs receive blood. Physicians measure blood pressure with an instrument called a sphygmomanometer.

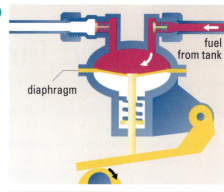

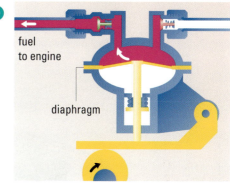

Figure 2

a In a car fuel pump, the diaphragm pulls down allowing fuel to enter the pump chamber.

b Fuel is pushed into the engine when the diaphragm pushes up.

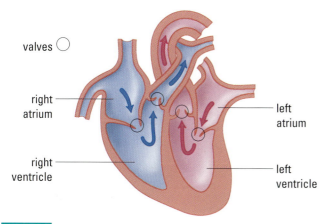

Figure 3
The heart is essentially a pump with valves that pushes blood around your body.

Design Challenge

What conductors will you use to move the fluid in your Design Challenge?

Understanding Concepts

1. List two industries that use hydraulic power.

2. What makes the fluid flow in a pressurized system? What controls this flow?

Making Connections

3. **(a)** What conductors can be used in a hydraulic system?

   (b) What conductors serve this function in the human circulatory system?

   (c) What conductors are found in a tree? What is the fluid that is being moved?

4. Compare and contrast a car fuel pump with a human heart using a Venn diagram.

Reflecting

5. How has knowledge of hydraulics aided our understanding of the human circulatory system?

Pressurized Fluid Systems: Pneumatics

Pneumatics is the name given to confined, pressurized systems that use moving air or other gases such as carbon dioxide. Like hydraulic systems, pneumatic systems possess a great deal of power that can be used to move an object.

air enters

cylinder

drill

piston

Figure 1

The jackhammer is a pneumatic drill. Compressed air moves a piston up and down, which moves the drill. These portable machines are often used to break apart concrete.

What Makes Up a Pneumatic System?

A pneumatic system is very similar to a hydraulic system (**Figure 1**). An air compressor provides the supply of air in a pneumatic system. The air compressor serves a similar purpose to the pump in a hydraulic system.

Pneumatics operate machinery such as air conditioning systems in aircraft and ejection seats in fighter planes. Pneumatic wrenches are used to remove or tighten nuts during a tire change. **Figure 2** shows a pneumatic drill in operation.

Figure 2

Pneumatic drills hammer away at concrete to break it up, ready for removal.

The Lungs: A Pneumatic System

Your lungs operate like a pump. They draw in air laden with oxygen and push out air and extra carbon dioxide (see **Figure 3**). They could not function without the diaphragm. The diaphragm works in a similar way to the plunger in a syringe. When the diaphragm pulls away from the lungs, the volume of the lungs increases, and the pressure inside lowers. The air outside your body is then at a higher pressure than the air in your lungs. This causes air to rush in. To expel gas, the diaphragm pushes up, the pressure inside the lungs increases, and you exhale. About 1L of air always remains in your lungs to prevent them from collapsing.

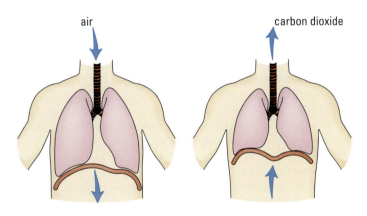

air

carbon dioxide

Figure 3

When we breathe in the diaphragm contracts; when we breathe out, the diaphragm relaxes.

Understanding Concepts

1. Using your diagrams and observation notes from the Try This, write a sentence summarizing the effects of a valve in a pneumatic or hydraulic system.

Making Connections

2. How does the gas on top of the liquid in an aerosol can cause the liquid to come out of the spray nozzle?

3. Compare and contrast a car fuel pump with a human lung using a Venn diagram.

Exploring

4. How does chewing gum relieve pressure inside your ears?

5. Research how natural gas is (4A) distributed in your community.

Design Challenge

For the safety hinge challenge, predict what would be different about the way a pneumatic and a hydraulic hinge would operate. Give reasons to support your prediction.

Try This   Exploring Valves

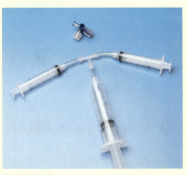

Figure 4

Be careful when working with fluids under pressure.

Find out how using a valve alters a pneumatic system.
- Add a valve to the pneumatic system you used in steps 4 to 7 of Investigation 2.18. The valve could go anywhere in the system.
- Draw your system.
- Move each of the plungers in turn and record your observations.
- Move the valve to another position in the system and repeat the procedure.
- Continue until you have tested all possible positions for the valve.

A Closer Look at Fluid Power

There are many kinds of fluid power systems all around us. At an airport, for example, fluid systems are used for passenger movement and baggage handling as well as control of aircraft systems such as doors, wheels, rudder and flaps (**Figure 1**). At the hairdresser's, the client's chair is moved up and down by fluid pressure (**Figure 2**). Even very large, heavy objects can be moved with the assistance of fluid-filled systems (**Figure 3**).

Materials
- apron
- safety goggles
- support stand
- screw-on clamp
- 20-mL syringe, 2
- 5-mL syringe
- 2 one-way valves
- 40 cm of clear 6-mm tubing, plus several shorter pieces
- water
- plastic container or beaker with a pour spout
- 500-g ball of modelling clay
- sponge

(3B) Problem
Reread the introductory paragraph. Think of a need for a fluid power lift. You have been hired by a management company to design this lift.

Design Brief
Design a hydraulic or pneumatic system that will raise or lower objects. Build a model of your design for test purposes and to present to the management company.

Design Criteria
- The model must raise a mass of 500 g to a specified height of 6 cm, remain stationary for at least 30 s, then descend in a controlled manner.
- The model must use only the materials listed.

Build

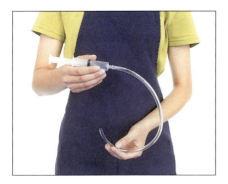

 (a) Record challenges or problems that arise during the design and construction of your model.

(b) Draw your completed model. Include the following labels on your diagram: cylinder, piston, and conductor.

(3D) **1** Design your model lift.
- After your teacher has
(3E) approved your design, build your model.

Test

2 How well does your model meet the design criteria?

(3F) (a) If not all of the design criteria are met, what changes do you need to make to your design?

- Make the necessary changes.

 Water-filled syringes can be quite dangerous when under pressure. Check your connections carefully first.

Figure 1

Figure 2

Figure 3

At the Big Chute Marine Railway on the Trent-Severn Waterway, hydraulic cylinders operate slings that support boats as they are carried over a 17.7-m height of land.

Evaluate

3 Evaluate your results by answering the following questions.

(a) When is the fluid in your system being compressed?

(b) Why must no air be present in a system filled with water?

(c) How would you notice if your air-filled system were leaking?

Design Challenge

How has designing and building a model fluid power system assisted you in designing your Challenge?

Making Connections

1. What difference might you notice if your hydraulic model were filled with oil?

Exploring

2. How would you modify your model to lift a load twice as heavy?

3. In your fluid system, clay was used as the object to be lifted. What changes would you need to make to your system to lift a stiff, rectangular object?

4. Set up your lift as a hydraulic system. Exchange the syringe providing the effort force with a smaller syringe. Describe the change in the effort now required to lift the load on the larger syringe.

Reflecting

5. What are two benefits of hydraulic and pneumatic lifts?

Fluid Power at Work for Us

Hydraulic and pneumatic systems are versatile. They can be used to do very heavy or extremely delicate work. Tiny hand-held drills operated by pneumatics are used for medical surgery. Hydraulics and pneumatic robots prevent human injury by performing dangerous jobs on assembly lines. Fluid power machines save industries money, doing heavy tasks quickly and efficiently that would take many people long hours to perform. Hydraulic or pneumatic systems are combined with electrical systems and solid mechanical systems (pulleys and levers) in an amazing variety of ways to meet the needs of society.

Working to Entertain Us

From special effects that include huge moving beasts (**Figure 1**) to amusement park rides (**Figure 2**), fluid power systems work to frighten and thrill us.

Figure 2

A pneumatic system is used to slow or stop roller coasters.

Figure 1

Animated movie figures come to life because of hydraulic systems.

Hydraulics to the Rescue

There are many kinds of Jaws of Life tools. (See **Figure 3**.) Some can cut with a force as high as 169 kN. Others are used to pry things apart. These hydraulic devices are at work opening the sides of cars and slicing guardrails at the roadside to get accident victims to the hospital quickly.

Figure 3

The Jaws of Life tool is a hydraulic cutting-and-prying machine. It uses a hydraulic fluid made especially for accident scenes where the risk of fire or explosion is high. The fluid is fire resistant and does not conduct electricity.

Training Uses

Figure 4

Hydraulic systems are used to create motion in flight and driving simulators. (See **Figure 4**.) Operators sit inside a model of the real vehicle and respond to computer-generated situations as if they were real. Hydraulic cylinders move the whole simulator back and forth and from side to side. Because this is a simulation, dangerous manoeuvres can be tried without anyone getting hurt. Hydraulics is at work to give us the best-trained pilots and drivers.

Moving Earth Beneath Our Feet

Figure 5 illustrates a section of a Tunnel Boring Machine (TBM) at work building a Toronto subway tunnel. The machine has two functions: it bores through the earth to form the tunnel, and it erects the lining of the tunnel.

To form the tunnel, 14 hydraulic motors rotate the cutting head that excavates the ground. While the cutting head digs, 24 hydraulic-thrust cylinders push the machine forward. The excavated soil passes through hydraulically operated doors to a screw-type conveyor. A second conveyor belt takes this soil to waiting rail cars, which haul the dirt away.

As the tunnel is being formed, the lining is built. The concrete tunnel lining is pre-built in segments, which are put in position by a machine and bolted together using pneumatic wrenches.

Boring 1m of tunnel an hour, the fluid power systems in this modern machine are underground working for us.

Understanding Concepts

1. List three benefits of fluid power systems.
2. Compare and contrast hydraulics and pneumatics.

Making Connections

3. What amusement park ride uses hydraulic systems to create motion similar to a flight or driving simulator?

Exploring

4. Make a list of ten devices or machines that use fluid power. State whether each is a hydraulic or pneumatic system.

Reflecting

5. Why is it important to learn about fluids and fluid power systems?

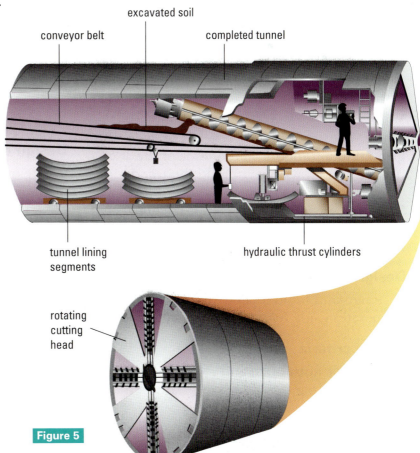

excavated soil
conveyor belt
completed tunnel
tunnel lining segments
hydraulic thrust cylinders
rotating cutting head

Figure 5

Design Challenge

SKILLS MENU
- Identify a Problem
- Planning
- Building
- Testing
- Recording
- Evaluating
- Communicating

Design and Build a Device That Uses the Properties of Fluids

Fluids are a vital part of our natural world. For centuries we have been trying to learn more about the principles of fluids. As our understanding has increased, so too has our use of fluids. Today, the applications of these fluid principles are numerous and have had a great impact on society. By designing and building a fluid device, you will gain a better understanding of the use of fluids.

1   A Safety Hinge

This door operates by pneumatic fluid power.

Problem situation

You are helping some friends build a tree house. The access ladder reaches up from the ground to a trapdoor in the floor of the platform. It is awkward to hold the trapdoor open as you climb through. You wish it would stay open, in an upright position, to allow you to climb in. You'd also like it to close gently, rather than slam down.

Design brief

- Design and build a safety hinge that operates using hydraulics or pneumatics.

Design criteria

- The door must open and close safely.
- After opening, the door must stay open for 15 s.

2   A Boat Navigation Lock

Problem situation

You are working with a team of engineers who are building a waterway between two lakes so that recreational boaters can travel from one to the other. The lakes are close together, but one is 5 m higher than the other. You don't want a rushing river between the two lakes, so you need some way to move the boats from one level canal to the other: you need a "lock."

Design brief

- Design and build a model of a lock that allows boats to travel both ways between waterways of different levels.

Design criteria

- The model lock must move a toy boat from one water level to the other, at least 5 cm higher or lower.

In 1998, the lock at Sault Ste. Marie, Ontario, was reopened after being closed for 11 years. The lock was originally built in the late 1800s to link the St. Lawrence Seaway with Lake Superior. Today, it is used as a recreational lock.

Process

- understand the problem
- develop a safe plan
- choose and safely use appropriate materials, tools, and equipment
- test and record results
- evaluate your model, including suggestions for improvement

Communicate

- prepare a presentation
- use correct terms
- write clear descriptions of the steps you took in building and testing your model
- explain clearly how your model solves the problem
- make an accurate technical drawing for your model

Produce

- meet the design criteria with your model
- use your chosen materials effectively
- construct your model
- solve the identified problem

3   A Fish-Tank Cleaner

Problem situation

Your Saturday job at the pet store involves scraping the green algae off the insides of the fish tanks. This is a messy job, and you wish there was some way of doing it that didn't involve getting up to your elbows in water. You decide to use your knowledge of buoyancy to build something that will help you.

Design brief

- Design and build a fish-tank cleaning device that will raise or lower itself in water.

Design criteria

- The fish-tank cleaner must be equipped with a scraper that cleans the walls of the fish tank.
- It must go up and down in the water as a result of its changing buoyancy.

 When preparing to build or test a design, have your plan approved by your teacher before you begin.

Unit 2 Summary

Now that you have completed the unit, can you do the following? If not, review the sections indicated.

Reflecting

- Reflect on the ideas and questions presented in the Unit Overview and in the Getting Started. How can you connect what you have done and learned in this unit with those ideas and questions? (To review, check the sections indicated in this Summary.)
- Revise your answers to the Reflecting questions in ❶, ❷, ❸ and the questions you created in the Getting Started. How has your thinking changed?
- What new questions do you have? How will you answer them?

Understanding Concepts

- explain, using examples, how fluid flow can be classified 2.2
- describe each of the three properties of fluids (viscosity, density, and buoyancy) you have investigated 2.3, 2.8, 2.12
- identify situations where understanding each property is important 2.5, 2.11, 2.12, 2.13, 2.14, 2.15, 2.16, 2.17
- describe how the forces of buoyancy and gravity act on objects immersed in fluids 2.11, 2.12

- distinguish between fluids and solids, using the particle theory of matter, and explain how changes in temperature affect their densities 2.1, 2.16
- define the terms hydraulics and pneumatics 2.20, 2.21

Applying Skills

- predict and measure the effect of temperature on the flow rate of fluids 2.4, 2.16

- calculate the mass-to-volume ratio of a substance 2.6, 2.7

- compare densities of fluids and solids to explain why some substances float on top of other substances 2.9, 2.10

- design, build, calibrate, and use a hydrometer 2.13

- describe what happens to fluids when they are under pressure 2.18, 2.22

- design and build a hydraulic or pneumatic system. 2.22

- understand and use the following terms:

> aerodynamics
> Archimedes' principle
> ballast
> buoyancy
> compressible
> cylinder
> density
> displacement
> drag
> dynamic
> flow rate
> hydraulic fluid
> hydraulics
> hydrodynamics
> hydrometer
> laminar flow
> mass
> negative buoyancy
> neutral buoyancy
> particle theory
> piston
> pneumatics
> positive buoyancy
> pressure
> streamlined
> turbulent flow
> viscometer
> viscosity
> volume
> weight

Making Connections

- describe situations where the use and investigation of fluids have affected our lives 2.7, 2.12, 2.14, 2.15, 2.19, 2.20, 2.21, 2.22, 2.23

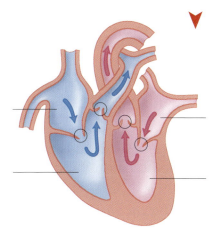

- explain how changes in fluids' viscosity and density can be useful 2.3, 2.4, 2.5, 2.10, 2.12, 2.14, 2.16

- identify a career that requires an understanding of the properties of fluids 2.5, 2.12, 2.17

- describe real life situations where scientists use their knowledge and understanding of fluids to solve challenges 2.5, 2.17

- discuss the versatility of hydraulic and pneumatic systems 2.23

Unit 2 Review

Understanding Concepts

1. Make a list of five fluids that can be found in each of these places:
   (a) the human body
   (b) a kitchen
   (c) a garage

2. Make a list of ten machines or devices that use fluid power systems.

3. Identify five industries where the properties of fluids play an important role. For each industry, provide an example of fluid use.

4. Using the pictures and statements provided in the Unit Summary, develop a concept map of the *Fluids* unit.

5. The following statements contain information about mass or weight. In your notebook, write the letters "a" to "f." Next to each letter, write the word "mass" if that letter's statement refers to the measurement of mass. Use the word "weight" if the statement refers to a weight measurement.
   (a) This measures the amount of matter in an object.
   (b) This measures the force of gravity acting on an object.
   (c) This measurement varies according to the location of the object in the universe.
   (d) Unless something is added to or taken away from the object, this measurement of an object remains the same everywhere in the universe.
   (e) This pulling force is measured in newtons.
   (f) This measurement is not a force.

6. Density is a property of fluids and solids.
   (a) What is the meaning of the term density?
   (b) Why is the density of a substance compared to the density of water?

7. (a) Create a poem about bouyancy, showing that you understand what it means.
   (b) Using the term "buoyant force," explain why a life jacket keeps a person afloat. (See **Figure 1**.)

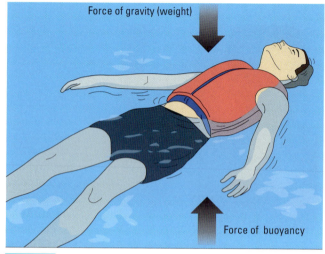
Force of gravity (weight)

Force of buoyancy

Figure 1

8. Compare how a submarine, fish, and scuba diver control their depth in the water.

9. Using the particle theory, explain the effects of temperature changes on solids, liquids, and gases. Draw diagrams to support your explanation.

10. Fill in the blanks in the following sentences with the correct words, whose letters are found scrambled in parentheses.
    (a) The _____?_____ of a fluid is an indicator of its viscosity. (lwof eatr)
    (b) When the flow of fluid around an object is smooth and uniform, it is referred to as _____?_____ flow. (ranamil)
    (c) The _____?_____ of measurement for density is g/cm3 or kg/m3. (itnu)
    (d) Because the volume of a liquid under pressure changes very little, the liquid is said to be virtually _____?_____. (biipmnosesrcel)
    (e) Helium balloons float in air because their _____?_____ is less than that of air. (yesidtn)

(f) Mechanical systems that use fluids to transmit force and move objects are called hydraulic or _____?_____ depending on the fluid used. (mtacipeun)

(g) The buoyant force _____?_____ the force of gravity. (spspooe)

(h) The force that measures the amount of gravity acting on a mass is _____?_____. (tewhig)

(i) A living thing not normally found in the area where it is living and reproducing is said to be an _____?_____. (toecix eesscip)

(j) Streamlined shapes produce less drag or _____?_____ when they are moving through fluids. (cratesiens)

(k) The first letter from the first word in each blank above spells the answer to the following question: What do you get when confined fluids operate under pressure?

11. Look at the three sketches in **Figure 2** and answer the following questions:

(a) What measurement is being taken of the rock in air, water, and space?

(b) How does the mass of the rock change at the three locations?

(c) How does the weight of the rock compare among the three locations?

(d) How does the buoyant force acting on the rock differ in air and water?

(e) In what location is the force of gravity the smallest?

(f) Does the force of gravity acting on the rock differ between the air and water locations? Explain.

Applying Skills

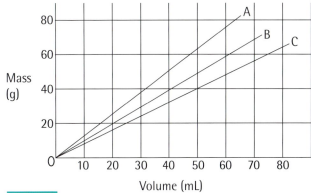

Figure 3

In air on Earth   In water on Earth   In space away from Earth

Figure 2

12. The graph in **Figure 3** illustrates the relationship between mass and volume for three liquids, A, B, and C. Using the information in this graph, answer the following questions:

(a) Which liquid has the highest density?

(b) Which liquid has the lowest density?

(c) Suppose you took these three liquids and poured them slowly, one at a time, into a tall container. Draw a diagram of where you would find each liquid in the container.

(d) Calculate the density of liquid B. What could be the identity of this liquid?

13. Your teacher has asked you to design an investigation in which you use an eyedropper to measure the flow rate of the three liquids in question 12.

(a) Describe how you would do this investigation.

(b) What variables will you have to keep constant as you test each liquid?

(c) What sources of experimental error might you encounter?

(d) From the information in the graph for question 13, which substance might you think has the highest flow rate?

(e) Could your prediction in (d) be proven incorrect when you perform the investigation? Explain.

14. The reading on a hydrometer standing upright in a liquid is 1.24.

(a) Explain what this measurement represents.

(b) Would this liquid float on water? Explain.

(c) Draw a diagram of the above liquid containing the hydrometer, beside the same hydrometer in a less dense liquid.

15. Which diagram in **Figure 4** represents the particles in a solid? In a liquid? In a gas?

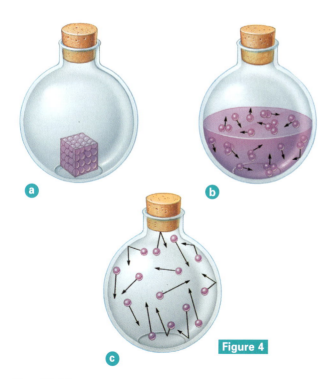

Figure 4

Making Connections

Figure 5

16. Laminar airflow provides a cleaner environment for this microbiologist working with microorganisms (**Figure 5**).

(a) Draw a picture that shows the invisible airflow inside the hood where the microbiologist is working.

(b) How does laminar airflow compare to turbulent flow?

(c) Describe two situations when laminar airflow around an object would be desirable.

17. In order to lubricate a car engine, engine oil must remain viscous. The "W" in motor oil stands for weight. 10W30 motor oil has the viscosity of lighter oil when it is cold, and the viscosity of heavier oil when it is hot. Why is it important that motor oil has these characteristics in Canada's climate?

18. Jacques Cartier returned to France after his explorations in North America with samples of what he thought were gold and diamonds. These samples turned out to be "fool's gold" and quartz. Explain how you could use Archimedes' principle to show that Cartier's samples were not the real thing.

19. Suppose your class is having a pool party (**Figure 6**). What will happen to the level of the water if each person enters the pool at the same time? How could you calculate this change in water level?

20. Two bodies of water, the Dead Sea and Great Salt Lake, are much saltier than oceans. Using the data from the table of densities in 2.8, explain why it is easy for a person to float in these locations.

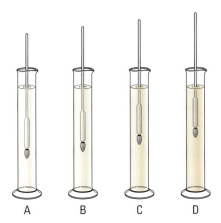

21. During the production of a batch of maple syrup, a hydrometer is placed in four test samples taken at different times throughout the evaporation process. Refer to the illustrations above.

  (a) Rank the liquids in order by density from least dense to most dense.

  (b) Which sample was collected earliest in the evaporating process?

  (c) Which sample was collected latest in the evaporating process?

  (d) Which sample would taste the sweetest?

22. Divers (and firefighters) carry a supply of air to breathe. Apply your knowledge from investigating fluids to explain how divers can remain under water for long periods of time with only a small tank of air.

23. Bicycle pumps push air into tires. What is the purpose of the valve on the end of the hose? (See **Figure 7**.)

24. Look at the picture of a car jack raising a car (**Figure 8**). Sketch the hydraulic cylinder inside the car jack.

25. A warning on an aerosol can states, "Caution, container may explode if heated." Explain, using the particle theory of matter, why such a warning is necessary.

26. Describe how the air pressure changes inside a soccer ball when it is kicked.

Glossary

A

aerodynamics: concerns the flow of air around solid objects or the effect of air on objects moving through it

Archimedes' principle: a rule that says the buoyant force on an object is equal to the weight of the fluid that the object displaces

B

ballast: any substance that acts as a weight and alters buoyancy of a vehicle, such as a ship, submarine, hot-air balloon, or dirigible

buoyancy: the upward force that a fluid exerts on an object less dense than itself

C

compressible: the ease with which a substance reduces its volume under external force

cylinder: a cylindrical chamber in a hydraulic system, which houses a piston that moves under fluid pressure

D

density: the mass of a substance per unit volume of that substance, calculated by dividing the mass of a substance by its volume

displacement: a technique used to measure the volume of small irregular solids, using the formula: volume of object = (volume of water + object) − (volume of water)

drag: a force that acts to slow a body moving through a liquid

dynamic: relating to systems involving moving fluids

F

flow rate: the speed that a fluid moves in a given amount of time

H

hydraulic fluid: a liquid under pressure in a hydraulic system that enables the system to work

hydraulics: confined, pressurized systems that use moving liquids to operate

hydrodynamics: the motion of liquids, usually water, around solid objects

hydrometer: an instrument used to measure the density of liquids

L

laminar flow: the movement of water in straight or almost straight lines

M

mass: the amount of matter in an object, measured in milligrams (mg), grams (g), or kilograms (kg)

N

negative buoyancy: the tendency of an object to sink in a fluid because the object weighs more than the fluid it displaces

neutral buoyancy: the tendency of an object to remain level in a fluid because the object weighs the same as the fluid it displaces

P

particle theory: a theory used to explain matter and heat transfer, which suggests that all matter is made up of tiny particles too small to be seen. These particles are constantly in motion because they have energy. The more energy they have the faster they move.

piston: a cylinder or disk inside a larger cylinder that moves under fluid pressure

pneumatics: confined, pressurized systems that use moving air or other gases

positive buoyancy: the tendency of an object to rise in a fluid because the object weighs less than the fluid it displaces

pressure: a measure of the amount of force acting on a certain area of surface

S

streamlined: a smoothly curved, narrow shape that allows an object to move easily through a fluid, disturbing the fluid as little as possible

T

turbulent flow: a broken or choppy movement of water usually caused by rapids, eddies, or whirlpools

V

viscometer: an instrument that measures the resistance of a fluid to flow and movement

viscosity: the physical property of a liquid that limits its ability to flow

viscous: having relatively high resistance to flow

volume: a measure of the amount of space occupied by matter, measured in cubic meters (m^3), litres (L), cubic centimeters (cm^3), or millilitres (mL)

W

weight: a measure of the force of gravity pulling on an object, measured in Newtons (N)

Index